Solid Waste Disposal and Reuse in the United States

Volume I

Editor

Ishwar P. Murarka, Ph.D., CPSS
Program Manager
Land and Water Quality Studies
Electric Power Research Institute
Palo Alto, California

CRC Press, Inc.
Boca Raton, Florida

Library of Congress Cataloging-in-Publication Data

Solid waste disposal and reuse in the United States.

Includes index.
1. Factory and trade waste--United States. 2. Refuse and refuse disposal--United States. 3. Recycling (Waste, etc.)--United States. I. Murarka, Ishwar P.
TD897.7.S65 1987 628.4′4′0973 86-24343
ISBN 0-8493-4647-9 (v.1)
ISBN 0-8493-4648-7 (v.2)

This book represents information obtained from authentic and highly regarded sources. Reprinted material is quoted with permission, and sources are indicated. A wide variety of references are listed. Every reasonable effort has been made to give reliable data and information, but the author and the publisher cannot assume responsibility for the validity of all materials or for the consequences of their use.

Direct all inquiries to CRC Press, Inc., 2000 Corporate Blvd., N.W., Boca Raton, Florida, 33431.

International Standard Book Number 0-8493-4647-9 (Volume I)
International Standard Book Number 0-8493-4648-7 (Volume II)

Library of Congress Card Number 86-24343
Printed in the United States

PREFACE

It has been both a privilege and a challenge to have undertaken the writing and editing of these volumes on solid waste. The purpose of these volumes is to provide an easy-to-use reference on solid waste disposal and reuse to environmental consultants, state and federal regulators, environmental managers, and the general public who deal with residues from industrial processes.

These two volumes bring together available data from a number of sources to provide a comprehensive view of trends in solid waste production, disposal, and reuse in the U.S. The first chapter is devoted to a general discussion of the various kinds of wastes produced in the U.S. The following chapters contain in-depth discussions of wastes from several major industrial-commercial-domestic groups: Chapter 2, manufacturing industries; Chapter 3, the electric utility industry; Volume II, Chapter 1, wastewater treatment sludges (sewage sludge); Chapter 2, foundry industries; Chapter 3, mining and milling industries, and Chapter 4, chemical industries. Each discussion covers types and quantities of wastes, physical-chemical characteristics, disposal systems, and known reuse practices.

Photographs, tables, and figures are shown when it is necessary to emphasize visual and numerical facts. These volumes contain useful appendices and some of the never-before-published information on solid waste. The editor of this book welcomes comments and suggestions from users of the text so that in the future, new books on the subject matter could be made to better meet the needs.

Ishwar P. Murarka
Palo Alto, California

THE EDITOR

Dr. Ishwar P. Murarka is the Program Manager for Land and Water Quality Studies, Environmental Science Department, Environment Division, for the Electric Power Research Institute (EPRI) based in Palo Alto, California.

Before joining the Institute in 1979, Dr. Murarka was a Scientist for five years with the Argonne National Laboratory, Argonne, Illinois. While there, he was responsible for research and analysis of environmental assessment problems relating to plant development. He also did statistical and mathematical modelings for aquatic effects of thermal power stations.

Previously, Dr. Murarka was with Texas Instruments Corporation for a year as a statistician. Based in New York, he worked on ecological studies on the Hudson River Estuary.

From 1971 to 1973, Dr. Murarka was at North Carolina State University doing postdoctoral studies on a National Institute of Health fellowship. He took his Ph.D. degree in 1971 and M.S. degree in 1968 from Oregon State University in Soil Science and Statistics. Dr. Murarka received an M.B.A. degree from the University of Chicago in 1979.

Originally from India, Dr. Murarka received an M.A. degree in 1964 in Geography and a B.A. degree with honors in Geography from the University of Calcutta.

Dr. Murarka is a member of the American Statistical Association, American Society of Agronomy, Soil Science Society of America, the Biometric Society, Sigma XI, and the American Society for Testing of Materials. Dr. Murarka has been a Certified Professional Soil Scientist (CPSS) since 1984.

CONTRIBUTORS

Larry L. Boersma, Ph.D., CPAg/SS
Professor
Department of Soil Science
Oregon State University
Corvallis, Oregon

Russell H. Boyd, M.S.,
Regional Manager
Ebasco Services, Inc.
Norcross, Georgia

H. P. Harbert, Ph.D., CPSS
Consultant
Forney, Texas

Nancy Hendrickson, M.S.,
Engineer
Tetra Tech, Inc.
Lafayette, California

Thomas P. Kunes, P.E.
Executive Vice President
Residuals Management Technology, Inc.
Madison, Wisconsin

Benjamin J. Mason, Ph.D.
President
ETHURA
Grants Pass, Oregon

Ishwar P. Murarka, Ph.D., CPSS
Progam Manager
Environmental Assessment Department
Electric Power Research Institute
Palo Alto, California

Karen V. Summers, M.S.
Principal Hydrogeologist
Tetra Tech, Inc.
Lafeyette, California

TABLE OF CONTENTS

Volume I

Volume II

Chapter 1

OVERVIEW OF SOLID WASTES IN THE U.S.

I. Murarka

TABLE OF CONTENTS

I. BACKGROUND

Solid waste disposal affects virtually everyone in the U.S.; all the industrial, municipal, agricultural, and domestic sectors produce solid residues of one sort or another. As late as the early nineteenth century, it was common practice to simply dump refuse in the streets. It was not until the mid-nineteenth century that epidemics and plagues were connected with this casual disposal and that collection and disposal in open pits was established. Since the turn of the century there has been a rapid expansion of the variety of solid wastes generated, due, in large part, to increased industrial activity coupled with increased environmental awareness. This led to the development of variations on the open-pit disposal method. Sanitary landfills, lagoons, ponds, and isolated engineered disposal systems are now being used for solid residues disposal.

Legislative attempts to regulate solid waste collection and disposal date back to 1899 when the Rivers and Harbors Act began regulating the dumping of debris in navigable waters and adjoining lands. However, it was not until 1965 that the first Solid Waste Disposal Act (PL89-272) was specifically enacted to regulate solid waste disposal. The Federal Resource Recovery Act (PL91-512), enacted in 1970 to amend the Solid Waste Disposal Act, shifted the objective from disposal of solid waste to recycling, resource recovery, and energy conversion of wastes. In 1976, the Federal Resource Conservation and Recovery Act (RCRA, PL94-580) replaced the previous laws and established a comprehensive role for the U.S. Environmental Protection Agency in the management and disposal of solid waste.

The principal aim of the RCRA is to encourage proper disposal of solid residues so that human health and the environment are not harmed and so that any usable resources within the wastes are recovered. The law recognizes that the ultimate success of the U.S. industrial society depends largely on the proper disposal and reuse of our solid residues. By encouraging resource recovery and the reuse of solid wastes, the law also leads us to reduce the volume of residues needing disposal. The RCRA was further amended in 1980 (PL96-482) and 1984 (PL98-616), with increased commitment to appropriate solid waste disposal and resource recovery.

II. WASTE TYPES

There are three forms which describe the variety of wastes from all sources. A waste can be in liquid, gaseous, or solid in form. However, these three forms are closely related by a conversion cycle. Once a liquid waste is generated, it can be transformed to a solid waste form. The same holds true for the gaseous wastes. Solid wastes when incinerated may result in the production of gaseous and liquid wastes in turn (Figure 1). In this book, the focus is on the solid wastes even though some of the intermediate liquid and gaseous forms may be discussed to describe the origin of the solid wastes.

III. DEFINITION OF SOLID WASTES

There is no single universally accepted definition of solid wastes. Solid discards, sludges, and wastewater may or may not be included. Generally, we think of residuals (e.g., combustion ashes and tailings materials, which are mostly solids and are to be thrown away) as solid wastes. However, there are more encompassing definitions that reflect the great variety of solid wastes produced in the U.S. For example, water-solid mixtures (sludges) containing as low as 2% solids by weight have been discussed as solid wastes. In this book, we will focus on solid wastes on a dry weight basis even though the initial wastes may be mixtures (e.g., the sludges mentioned above). This

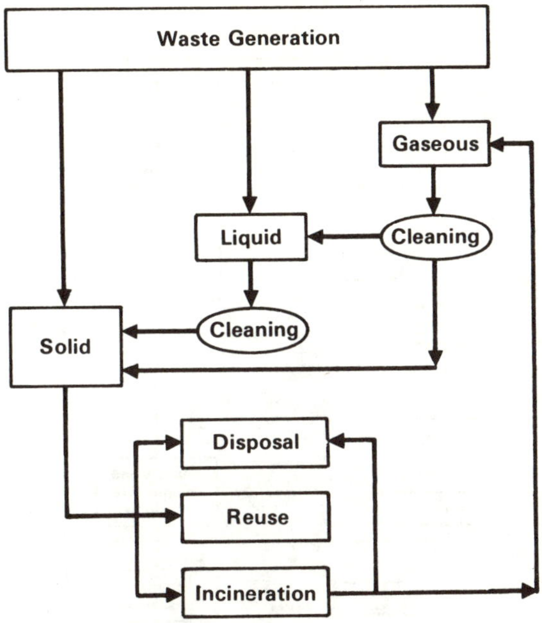

FIGURE 1. Waste types and their interrelationship.

approach will enable us to focus on disposal and reuse at the end point of the waste cycle.

IV. SOURCES AND QUANTITIES OF WASTES

Households, commerce, industries, agriculture, mining, and energy-related activities are all sources of solid wastes (see Figure 2). With the increase in technological dependence, solid discards are correspondingly increasing. Table 1 gives estimates of the quantitites of solid wastes generated in the U.S. in 1977.[1-3] The agriculture and mining-milling sectors are by far the largest generators of solid wastes, accounting for over 90% of the total annual rates of production. The industrial, municipal, and utility sectors account for the remaining 10%.

Waste generated by the residential and commercial sectors is quite varied in nature (see Table 2). The other sectors produce more uniform residues; e.g., the wastes from mining and benefication of ores are essentially rock and earthlike materials. Agricultural wastes are residues from plants, livestock, logging and timbering: corn, soybeans, and wheat account for the largest portion of crop residues, mostly straw; livestock operations produce animal wastes which are then used as manure; and the timber and logging industry generates solid wastes in the form of unusable rejects from the trees, including bark, leaves, and small branches.

V. SOLID WASTE DISPOSAL PRACTICES

In general, solid wastes are collected and then disposed of onto the land at disposal facilities that vary in size and configuration. The facility may be a secured or unsecured landfill, a surface pond, an impoundment, or a lagoon. It may cover less than an acre or several hundred acres. It may simply be a low-lying area into which the waste is dumped with no special preparation, or it may be highly engineered and specially lined with clay or polyethylene. The landfill may be a public site, a municipal operation

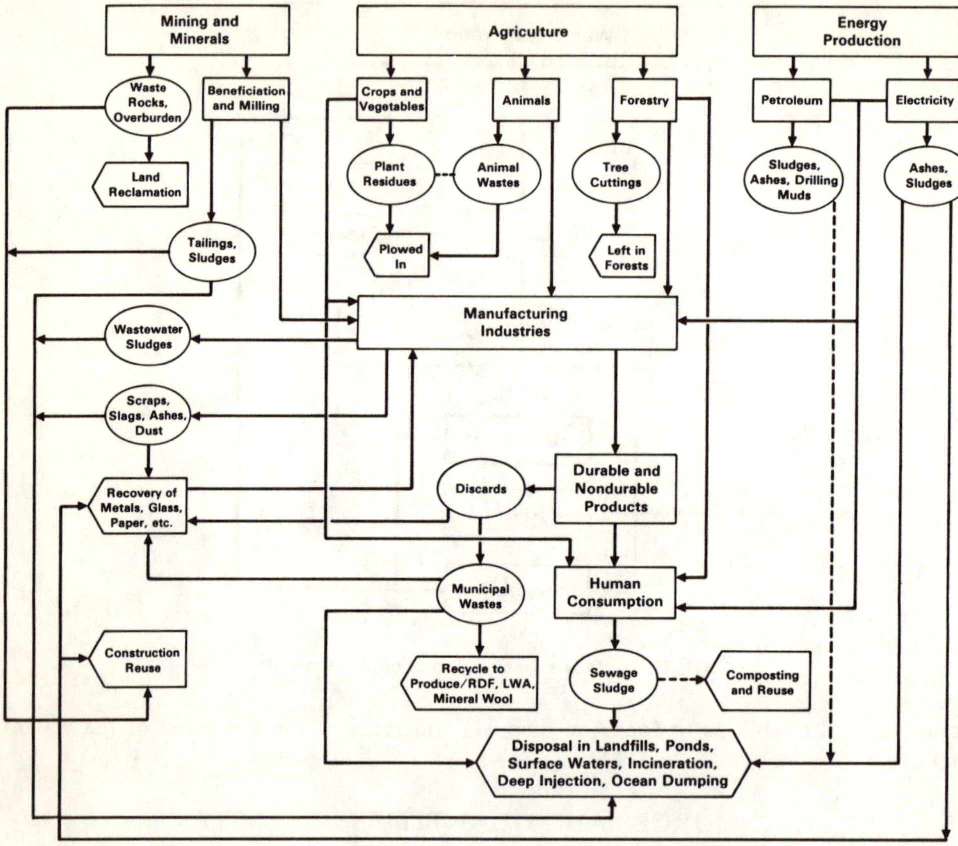

FIGURE 2. Sources and types of solid waste generation, disposal and reuse practices.

Table 1
ESTIMATED QUANTITIES OF SOLID WASTES GENERATED IN U.S.[1]

Waste source	1977[a]
Municipal	132
Sewage sludge	4.5
Junked autos and construction demolition	41
Industrial	326—362
Mining and milling	2086
Agricultural	2265—3014
Utility	70
Total	4925—5710

[a] Millions metric tons.

reserved for one city's residential and commercial waste disposal, or a private industrial site. Finally, the facility may be adjacent to or far removed from the site of waste generation. In addition to land disposal, incineration and ocean disposal are also used.

Landfills are commonly constructed for the specific purpose of disposal of the solid wastes. Ponds, on the other hand, are used for temporary storage or treatment or permanent disposal purposes. Many ponds are percolation/evaporation facilities for

Table 2
COMPONENTS OF MUNICIPAL SOLID WASTES COLLECTED FROM RESIDENTIAL AND COMMERCIAL SECTORS[2]

Component type	% of Total waste
Paper	31.6
Glass	9.8
Metals	9.4
Ferrous	(8.4)
Aluminum	(0.7)
Other nonferrous	(0.3)
Plastics	3.5
Rubber and leather	2.7
Textiles	1.5
Wood	3.7
Food waste	17.2
Yard waste	19.2
Miscellaneous wastes	1.4
Total	100.0%

the explicit purpose of allowing water to seep into the ground or evaporate into the atmosphere, leaving behind the dissolved constituents as sludge. Other ponds are specifically designed to retain the water and solids within the confines of the pond.

VI. SOLID WASTE DISPOSAL FACILITIES

Aside from the complex specifics of the various types of disposal facilities, it is not even known how many land disposal facilities exist and how big they are. According to SCS Engineers report,[4] there were some 93,384 landfills in the U.S. as of December 1980 (Table 3);[4,5] 77,807 were industrial landfills and 15,577 were municipal facilities (see Table 4 for capacities, depth, and areas covered). About 23,000 of these landfills receive municipal sewage sludge.[3]

According to the report by Geraghty and Miller, Inc.,[6] there were some 132,709 surface impoundments in the U.S. as of 1978 (see Table 5). About half of these impoundments are associated with oil and gas extraction operations. Another fifth of these ponds are to be found at industrial sites.

Available estimates indicate that as high as 80% of industrial wastes are disposed of on-site in nonsecure landfills, ponds, and lagoons. Mining and milling industry wastes are usually deposited adjacent to the facility mostly as mill-tailings ponds. The utility industry has about 400 coal-burning power plants, each with at least one solid waste disposal facility — pond or landfill. Manufacturing industries use surface impoundments for treating and disposing of liquid wastes and sludges. Agricultural solid wastes from farming, logging, and animal production are broadly distributed. Although much crop residue is turned under the soil or fed to livestock, much logging and timber residue is left in the forests, and livestock manure is commonly spread on the land. Small amounts of these solid wastes find their way to the municipal landfills or end up in the wastewater treatment plants where they get incorporated in the sewage sludge produced.

Table 3
NUMBER OF MUNICIPAL AND INDUSTRIAL LANDFILLS BY STATE AS OF DECEMBER 1980[4]

State	Grand total[a]	Industrial total	Municipal total	Municipal inactive total[b]	Municipal active: Publicly owned	Municipal active: Privately owned
Alabama	1,295	1,150	145	10	115	20
Alaska	274	74	200	74	74	52
Arizona	601	469	132	12	108	12
Arkansas	962	652	310	60	179	71
California	10,500	10,000	500	50	225	225
Colorado	853	638	215	15	150	50
Connecticut	778	638	140	70	65	5
Delaware	27	14	13	2	11	—
District of Columbia	6	0	6	3	2	1
Florida	2,722	2,210	512	300	106	105
Georgia	2,085	1,694	391	79	281	31
Hawaii	180	151	29	1	24	4
Idaho	385	275	110	20	80	10
Illinois	4,930	4,580	350	70	205	75
Indiana	2,140	1,890	250	50	50	150
Iowa	1,065	805	260	60	143	57
Kansas	885	691	194	14	133	47
Kentucky	958	723	235	35	91	109
Louisiana	1,295	845	450	50	355	45
Maine	812	432	380	30	250	100
Maryland	884	757	127	72	45	10
Massachusetts	2,787	2,497	290	4	238	48
Michigan	5,800	4,400	1,400	400	600	400
Minnesota	1,557	1,307	250	50	100	100
Mississippi	853	608	245	45	170	30
Missouri	1,634	1,514	120	—	48	72
Montana	426	201	225	—	213	12
Nebraska	886	286	600	63	360	177
Nevada	211	91	120	—	120	—
New Hampshire	423	313	110	30	70	10
New Jersey	3,865	3,625	240	40	100	100
New Mexico	811	211	600	100	100	400
New York	8,386	7,693	693	139	424	130
North Carolina	2,145	1,985	160	20	70	70
North Dakota	374	104	270	10	174	86
Ohio	4,728	4,488	240	10	97	133
Oklahoma	986	756	230	10	143	77
Oregon	237	111	126	24	78	24
Pennsylvania	4,733	4,368	365	45	124	196
Rhode Island	682	660	22	1	18	3
South Carolina	1,031	871	160	3	76	81
South Dakota	456	116	340	40	268	32
Tennessee	6,200	6,000	200	66	118	16
Texas	4,680	3,480	1,200	400	400	400
Utah	462	300	162	17	145	—
Vermont	290	190	100	30	60	10
Virginia	1,299	1,029	270	20	167	83
Washington	1,345	1,220	125	5	60	60
West Virgina	491	411	80	20	40	20
Wisconsin	1,750	200	1,550	300	1,100	150
Wyoming	219	84	135	15	60	60
Total	93,384	77,807	15,577	2,984	8,433	4,160
Average	1,868	1,556	311	60	167	83

[a] Industrial and municipal Landfills combined.

[b] Inactive for less than 10 years. Sites closed more than 10 years ago are not included.

Table 4
NUMBER, CAPACITIES, DEPTH, AND AREA FOR MUNICIPAL LANDFILLS BY STATE AS OF DECEMBER 1980[4]

	Number of municipal landfills[a]						
State	Total	Less than 500,000 tons	500,000 to 1,000,000 tons	1,000,000 to 2,000,000 tons	More than 2,000,000 tons	Average depth (ft)[b]	Average acreage (ac)[b]
Alabama	145	116	16	9	4	40	50
Alaska	200	197	2	1	—	40	30
Arizona	132	107	14	8	3	40	45
Arkansas	310	275	25	8	2	30	50
California	500	410	11	27	52	90	75
Colorado	215	175	30	6	4	40	45
Connecticut	140	118	15	6	1	35	50
Delaware	13	7	2	2	2	34	50
District of Columbia	6	—	4	2	—	35	50
Florida	512	452	40	15	5	20	60
Georgia	391	322	55	12	2	35	40
Hawaii	29	22	5	2	—	40	30
Idaho	110	101	8	1	—	30	35
Illinois	350	75	120	75	80	35	150
Indiana	250	220	9	11	10	33	81
Iowa	260	254	3	3	—	30	50
Kansas	194	167	21	3	3	40	50
Kentucky	235	150	50	29	6	35	50
Louisiana	450	433	8	6	3	20	80
Maine	380	375	3	2	—	30	40
Maryland	127	102	6	12	7	35	60
Massachusetts	290	232	43	12	3	30	60
Michigan	1,400	1,200	120	65	15	40	60
Minnesota	250	228	9	3	10	40	70
Mississippi	245	243	—	—	2	25	120
Missouri	120	107	3	6	4	30	75
Montana	225	220	3	2	—	40	30
Nebraska	600	592	3	3	2	30	70

Table 4 (continued)
NUMBER, CAPACITIES, DEPTH, AND AREA FOR MUNICIPAL LANDFILLS BY STATE AS OF DECEMBER 1980[4]

State	Number of municipal landfills[a] Total	Less than 500,000 tons	500,000 to 1,000,000 tons	1,000,000 to 2,000,000 tons	More than 2,000,000 tons	Average depth (ft)[b]	Average acreage (ac)[b]
Nevada	120	117	1	—	2	40	65
New Hampshire	110	105	3	2	—	40	30
New Jersey	240	205	20	12	3	30	60
New Mexico	600	590	7	2	1	30	30
New York	693	503	100	80	10	40	40
North Carolina	160	156	3	1	—	35	25
North Dakota	270	269	1	—	—	30	25
Ohio	240	81	75	70	14	40	45
Oklahoma	230	219	5	6	—	35	35
Oregon	126	111	9	—	6	40	60
Pennsylvania	365	244	33	68	20	45	40
Rhode Island	22	18	2	2	—	30	40
South Carolina	160	116	20	19	5	30	60
South Dakota	340	338	1	1	—	30	30
Tennessee	200	168	18	8	6	40	40
Texas	1,200	1,136	20	25	19	40	80
Utah	162	154	5	1	2	40	30
Vermont	100	98	2	—	—	30	30
Virginia	270	262	5	3	—	40	30
Washington	125	105	7	9	4	45	40
West Virginia	80	60	9	7	4	45	45
Wisconsin	1,550	1,525	14	7	4	50	40
Wyoming	135	133	2	—	—	35	25
Total	15,577	13,613	990	654	320		

[a] Both active and inactive municipal sites. Capacities shown are tonnage in-place for inactive sites, planned tonnage for active sites.
[b] Data based on both active and inactive municipal site greater than 500,000 tons size.

Table 5
NUMBER OF SURFACE IMPOUNDMENTS BY MAJOR CATEGORIES AND STATES AS OF 1978[6]

State	Grand total	Municipal	Private/ commercial	Institutional	Industrial	Oil and gas extraction	Agricultural
Alabama	1,590	156	130	50	583	2	669
Alaska	130	6	—	—	100	9	15
Arizona	332	52	63	55	44	6	112
Arkansas	953	180	50	10	66	540	107
California	3,721	245	560	118	782	910	1,106
Colorado	5,237	137	125	10	103	4,617	245
Connecticut	96	11	—	—	48	—	37
Delaware	63	4	10	3	33	—	13
Florida	2,035	118	1,200	150	217	0	350
Georgia	1,438	195	225	79	205	—	734
Hawaii	78	5	5	5	29	—	34
Idaho	584	76	20	10	24	—	454
Illinois	3,667	413	100	50	445	2,000	659
Indiana	2,538	183	117	62	357	752	1,067
Iowa	1,466	257	245	54	210	—	700
Kansas	6,086	256	64	14	164	4,525	1,063
Kentucky	2,141	29	26	6	944	1,000	136
Louisiana	9,997	81	150	50	552	8,841	323
Maine	237	3	10	5	44	—	175
Maryland	523	53	20	18	321	—	111
Massachusetts	73	5	15	5	40	—	8
Michigan	3,229	142	93	3	279	2,020	692
Minnesota	1,540	235	54	14	52	—	1,185
Mississippi	1,676	283	129	49	266	363	586
Missouri	2,757	332	1,092	50	213	6	1,064
Montana	1,363	130	20	8	64	825	316
Nebraska	2,329	223	15	8	1,180	200	703
Nevada	261	23	50	10	110	15	53
New Hampshire	105	8	10	2	41	—	44
New Jersey	277	9	20	5	230	—	13

Table 5 (continued)
NUMBER OF SURFACE IMPOUNDMENTS BY MAJOR CATEGORIES AND STATES AS OF 1978[6]

State	Grand total	Municipal	Private/ commercial	Institutional	Industrial	Oil and gas extraction	Agricultural
New Mexico	16,176	38	28	2	63	16,000	45
New York	960	33	100	50	308	265	204
North Carolina	1,038	147	81	25	282	—	503
North Dakota	2,784	250	10	5	76	1,900	543
Ohio	13,196	88	100	50	1,460	11,000	498
Oklahoma	2,006	260	50	15	354	989	338
Oregon	757	104	28	10	88	—	527
Pennsylvania	15,341	51	41	90	12,300	2,500	359
Rhode Island	30	2	10	5	12	—	1
South Carolina	911	255	191	58	81	—	326
South Dakota	650	192	10	5	34	30	379
Tennessee	776	58	30	34	215	100[a]	339
Texas	8,436	379	358	142	1,042	6,000	515
Utah	669	38	8[a]	4[a]	68	317	234
Vermont	329	12	9	2[a]	244	0	62
Virginia	2,116	90	51	77	1,409	100	389
Washington	1,045	91	16	11	255	—	672
West Virginia	2,803	48	83	6	1,631	1,000	35
Wisconsin	985	223	50	11	103	—	598
Wyoming	5,179	64	15	5	73	5,000	22
U.S. Total:	132,709	6,273	5,887	1,510	27,844	71,832	19,363

VII. SOLID WASTE REUSE PRACTICES

Because of the increased production of solid wastes needing disposal, and because of increasing legislative and regulatory attention given to protection of the environment, incentives exist to increase the reuse of solid wastes. A growing recognition of the resource value of solid wastes has also played a role in increasing reuse. This reuse takes many forms; for example, recovery of valuable minerals and metals (largely from mining, milling, and municipal wastes) is being emphasized whenever economically and technologically feasible. Use of refuse for energy recovery is growing; a few facilities already process municipal collections for burning in boilers to produce electricity.

Solid wastes have been historically used in construction as subbase, backfills, and embankments, and this use is continuing. Increasing attempts are being made to use some of the wastes in products such as concrete, cement, aggregates, mineral wool, and ceramic and tile production. Finally, recycling of scraps of paper and metals is becoming economically feasible and serves two purposes: (1) it reduces solid wastes needing disposal, and (2) it conserves the mining of ores for the production of metals.

REFERENCES

1. Lacomber, D. M., *An Overview of Solid Waste Generation in the United States,* LA-8172-MS, Los Alamos Scientific Laboratory, Los Alamos, N.M., 1979.
2. Fourth Report to Congress—Resource Recovery and Waste Reduction, Office of Solid Waste, U.S. Environmental Protection Agency, Washington, D.C., SW-600, 1977.
3. Solid Waste Facts, Office of Solid Waste, SW-694, U.S. Environment Protection Agency, Washington, D. C., 1978.
4. SCS Engineers, Impact of the Resource Conservation and Recovery Act on Landfill Gas Utilization, ANL/CNSV-TM-117, Argonne National Laboratory, Argonne, Il., 1983.
5. 1980 Land disposal survey, *Waste Age,* 12, 1981.
6. Geraghty and Miller, Inc., Executive Summary of the Report—Surface Impoundments and Their Effects on Groundwater Quality in the United States—A preliminary survey, Rep. No. EPA570/9-78-055, U.S. Environmental Protection Agency, Washington, D.C., 1978.

Chapter 2

SOLID WASTES IN MANUFACTURING INDUSTRIES

I. P. Murarka, K. Summers, and N. Hendrickson

TABLE OF CONTENTS

I. INTRODUCTION

Manufacturing industries produce many and varied products including food, paper, leather, and lumber. Consequently, these industries produce varying quantities and types of solid waste.

The U.S. Department of Commerce has developed standard industrial codes to allow for the tracking of activities by similar industries.[1] This chapter covers manufacturing industries which are included in SIC codes 20 to 39 (Table 1). Several of the more important manufacturing industries and the wastes generated are discussed individually in subsequent sections. Much of the solid waste generated by these industries is in the form of wastewater treatment sludges. Municipal wastewater treatment sludges are discussed in detail in Chapter 4. The foundry and chemical industries are more thoroughly discussed in Chapters 5 and 7, respectively.

II. OVERVIEW OF SOLID WASTE GENERATION FROM MANUFACTURING

Manufacturing industries generate large quantities of waste. Table 2 summarizes the percent of the total manufacturing solid waste generated by each industry in 1983 and describes the types of solid wastes generated. Currently, the chemicals and allied products industry (SIC 28) and primary metals industries (SIC 33) account for the largest fraction of waste generation. These two groups each generated about 20% of the total in 1983. The other relatively large generators of solid wastes are: food and kindred products group (SIC Code 20), paper and allied products group (SIC Code 26), and the stone, clay, glass products group (SIC Code 32).

Table 3 summarizes the annual solid waste generation for each industry between 1973 and 1983.[2-12] Table 4 breaks this data down by state.[2-12] These tables indicate a steady rise in solid waste generation from 1973 to 1979 peaking at about 164 million short tons in 1979 with a rapid decrease to 89 million short tons in 1983. Figure 1 illustrates this trend in solid waste generation from manufacturing. This trend is also prevalent among many of the individual manufacturing industries. For example, it is consistent with the stone, clay, and glass manufacturing industries (SIC 32) (Figure 2A) and the primary metals manufacturing industries (SIC 33) (Figure 2B). Not all industries, however, exhibit the above trend. Solid waste generation for the petroleum and coal industries (SIC 29) (Figure 2C) as well as the closely related rubber and miscellaneous plastic product manufacturing (SIC 30) (Figure 2D) peaked in the early 1980s. In the fabricated metal products industry (SIC 34) (Figure 2E), solid waste generation has been fairly steady at around 1.9 million short tons per year. Contrary to these latter trends, solid waste generation in paper and allied products (Figure 2F) as well as in printing and publishing (Figure 2G) has been increasing steadily. Most industries generated unusually low quantities of solid waste in 1982 due to an industry-wide recession.

Manufacturing solid waste generation does not always parallel manufacturing production, but generally maximum and minimum waste generation corresponds with maximum and minimum production. Table 5 summarizes the annual production index (as value-added weights) for each SIC code.[14,15] The data have been normalized to 1967.

The annual production index and solid waste generation data, both normalized to 1974, are plotted for the four largest solid waste generators: (1) the chemical and allied products industries, (2) the primary metals industries, (3) the food industry, and (4) the paper industry. Both the chemical (Figure 3A) and food industries (Figure 3B) show

Table 1
SIC CODE AND THE ASSOCIATED MANUFACTURING INDUSTRY[a]

SIC code number	Title
20	Food and kindred products
201	Meat products
202	Dairy products
203	Preserved fruits and vegetables
204	Grain mill products
205	Bakery products
206	Sugar and confectionary products
207	Fats and oils
208	Beverages
209	Miscellaneous foods and kindred
21	Tobacco manufacturers
22	Textile mill products
24	Lumber and wood products
242	Sawmills and planing mills
243	Millwork, plywood, and structural members
244	Wood containers
245	Wood buildings and mobile homes
25	Furniture and fixtures
26	Paper and allied products
261	Pulp mills
262	Paper mills
263	Paperboard mills
265	Paperboard containers and boxes
27	Printing and publishing
28	Chemicals and allied products
281	Industrial inorganic chemicals
282	Plastics materials and synthetics
283	Drugs
284	Soaps, cleaners, and toilet goods
285	Paints and allied products
286	Industrial organic chemicals
287	Agricultural chemicals
289	Miscellaneous chemical products
29	Petroleum and coal products
291	Petroleum refining
295	Paving and roofing materials
30	Rubber and miscellaneous plastic products
31	Leather and leather products
32	Stone, clay, and glass products
324	Cement, hydraulic
327	Concrete, gypsum, and plaster products
33	Primary metal industries
331	Blast furnace and basic steel products
332	Iron and steel foundries
333	Primary nonferrous metals
334	Secondary nonferrous metals
335	Nonferrous rolling and drawing
336	Nonferrous foundries
34	Fabricated metal products
35	Machinery, except electrical
36	Electric and electronic equipment
37	Transportation equipment
38	Instruments and related products
39	Miscellaneous manufacturing industries

[a] From U.S. Office of Management and Budget, Standard Industrial Classification Manual, U.S. Government Printing Office, Washington, D.C., 1972.

Table 2
PERCENT AND TYPE OF SOLID WASTE GENERATED BY MANUFACTURING INDUSTRY

SIC code	Manufacturing industry	Percentage of 1983 manufacturing solid waste[a]	Types of solid wastes[b]
20	Food and kindred products	10.8	Meats, fats, oils, bones, offal vegetables, fruits, nuts and shells, and cereals
21	Tobacco manufacturers	0.2	—
22	Textile mill products	0.7	Cloth and fiber residues
23	Apparel and other finished products	0.4[c]	Cloth, fibers, metals, plastics, and rubber
24	Lumber and wood products	4.5	Scrap wood, shavings, sawdust, metals, plastics, fibers, glues, sealers, paints, and solvents
25	Furniture and fixtures	0.3	Those listed under Code 24, cloth and padding residues, resins, glass, rubber, adhesives, and paper
26	Paper and allied products	15.3	Paper and fiber residues, chemicals, paper coatings, fillers, inks, glues, and fasteners
27	Printing and publishing	3.6	Paper, newsprint, card board, metals, chemicals, cloth, inks, and glues
28	Chemicals and allied products	21.1	Inorganic and organic chemicals, paints, solvents, rubber, glass oils, pigments, metals, and plastics
29	Petroleum and coal products	4.0	Asphalt and tars, felts, asbestos, paper, cloth, and fiber
30	Rubber and miscellaneous plastic products	1.7	Scrap rubber and plastics, lampblack, curing compounds, and dyes
31	Leather and leather products	0.3	Scrap leather, thread, dyes, oils, processing and curing compounds
32	Stone clay and glass products	6.9	Glass, cement, clay, ceramics, gypsum, asbestos, stone, paper, and abrasives
33	Primary metal industries	19.8	Ferrous and nonferrous metal scrap, slag, sand, cores, patterns, and bonding agents
34	Fabricated metals products	2.2	Metals, ceramics, sand, slag, scale, coatings, solvents, lubricants, and pickling liquors
35	Machinery, except electrical	2.0	Slag, sand, cores, metal scrap, wood, plastics, resins, rubber, cloth, paints, solvents, and petroleum products
36	Electric and electronic equipment	1.9	Metal scrap, carbon, glass, exotic metals, rubber, plastics, resins, fibers and cloth residues
37	Transportation equipment	3.4	Metal scrap, glass, fiber, wood, rubber, plastics, cloth, paints, solvents, and petroleum products

Table 2 (continued)
PERCENT AND TYPE OF SOLID WASTE GENERATED BY MANUFACTURING INDUSTRY

SIC code	Manufacturing industry	Percentage of 1983 manufacturing solid waste[a]	Types of solid wastes[b]
38	Instruments and related products	0.6	Metals, plastics, resins, glass, wood, rubber, fibers, and abrasives
39	Miscellaneous manufacturing industries	0.3	Metals, glass, plastics, resins, rubber, composition, bone, cloth, straw, adhesives, paints, and solvents

[a] U.S. Department of Commerce Bureau of Census, Pollution Abatement Costs and Expenditures 1983, MA-200(83)-1, 1985, 24.

[b] Mantell, C. L., *Solid Wastes: Origin, Collection, Processing, and Disposal,* John Wiley & Sons, New York, 1975, 1127.

[c] Estimated from Bureau of Census data.[12]

an initial increase in waste generation which parallels an increase in manufacturing production. This is followed by a marked decrease in solid waste generation despite a continuing increase in production. Waste generation in the paper industry (Figure 3C) and the primary metals industry (Figure 3D) closely parallels production.

As summarized in Table 3 and Figure 1, solid waste generation from manufacturing is decreasing with time. This decrease is partially related to a decrease in annual production as illustrated in Figure 3. However, this is not the only factor producing this decrease in the quantities of solid waste generated. Other possible factors include the increasing costs of solid waste disposal, the introduction and enforcement of more stringent waste disposal regulations proposed in the Resource Conservation and Recovery Act and Comprehensive Environmental Response Compensation and Liability Act (CERCLA), and the use of new treatment technologies such as improved sludge dewatering devices and incinerators.

III. FOOD AND KINDRED PRODUCTS (SIC Code 20)

A. Waste Generation

As more and more food items are prepared in the form of frozen, packaged, or canned goods, the residuals from food processing plants are increasing. It is estimated that 28% of the raw material entering the food manufacturing industry becomes waste.[13] Table 6 lists the quantities of solid wastes generated by state and year for this industry group.[2-12] Solid waste generated averages about 12 million tons per year with peak generation of 15 million tons in 1976. In 1983, the food and kindred products industry generated about 11% of the total manufacturing solid wastes. The food and kindred products industry is widely distributed and constitutes an important part of the U.S. economy. Table 6 indicates that California and Hawaii generate the largest quantity of solid waste from the food industry.

This industry processes raw materials primarily obtained from the agricultural sector. All aspects of the manufacturing of foods and beverages for human and pet consumption are included in this group. Examples include production of meat products, dairy products, fruits and vegetables, bakery products, sugar and confectionary products, oils, and beverages.

Table 3
ANNUAL QUANTITY OF SOLID WASTES GENERATED IN MANUFACTURING INDUSTRIES[2-12]

Name of industry	SIC Code	1973	1974	1975	1976	1977	1978	1979	1980	1981	1982	1983
All manufacturing industries		78,475	133,324	139,051	156,826	159,954	160,753	163,678	142,957	145,796	99,429	89,042
Food and kindred products	20	9,137	11,501	12,641	14,965	13,088	13,392	13,971	14,410	13,155	9,783	9,602
Meat products	201	781	1,239	1,024	1,316	1,256	1,620	1,791	2,495	1,811	544	646
Dairy products	202	371	499	506	425	905	1,045	896	843	783	739	733
Preserved fruits and vegetables	203	2,438	2,899	3,097	3,682	2,914	2,932	3,783	3,365	2,728	2,826	2,132
Grain mill products	204	649	745	708	751	804	760	889	905	1,070	930	592
Bakery products	205	284	418	1,026	1,031	270	343	512	217	303	266	390
Sugar and confectionary products	206	3,232	4,217	4,825	6,231	4,529	4,445	4,044	3,705	4,539	3,018	3,472
Fats and oils	207	269	202	259	302	234	—	274	277	277	179	283
Beverages	208	626	729	632	804	1,553	1,341	953	2,004	1,087	896	828
Miscellaneous foods and kindred	209	487	562	564	423	624	674	831	599	556	386	474
Tobacco manufacturers	21	113	148	150	164	136	182	148	157	174	190	222
Textile mill products	22	565	750	668	671	825	1,180	1,101	810	1,144	846	691
Lumber and wood products	24	4,407	6,887	8,143	9,328	6,310	6,662	6,506	5,893	6,419	3,747	3,994
Sawmills and planing mills	242	2,791	5,344	5,625	6,320	3,280	3,621	3,041	2,566	3,422	1,555	1,830
Millwork, plywood, and structural members	243	781	717	863	1,507	1,233	1,308	1,461	1,524	1,347	966	1,127
Wood containers	244	14	70	155	—	—	—	—	—	—	54	64
Wood buildings and mobile homes	245	103	86	95	100	103	126	385	—	—	61	55
Furniture and fixtures	25	422	517	592	638	668	705	736	668	621	346	296
Paper and allied products	26	6,709	8,837	9,123	10,146	10,648	10,947	13,253	12,251	11,299	11,452	13,655
Pulp mills	2,611	312	402	562	392	558	564	769	1,064	940	925	1,422
Paper mills	2,621	2,503	4,064	4,142	4,960	5,103	5,429	5,164	5,362	5,437	5,750	6,764
Paperboard mills	2,631	1,603	2,711	2,556	2,658	3,150	3,328	3,647	3,596	3,594	3,345	3,818
Paperboard containers and boxes	265	525	465	607	773	462	406	407	358	—	205	358
Printing and publishing	27	548	617	628	848	941	879	1,148	1,177	2,087	558	3,227
Chemicals and allied products	28	11,861	38,155	38,695	50,282	55,719	48,803	45,407	43,431	43,673	36,028	18,816
Industrial inorganic chemicals	281	3,161	7,177	7,772	8,254	9,170	8,363	9,018	12,528	11,752	7,175	5,929
Plastics materials and synthetics	282	1,070	1,548	1,640	1,968	1,647	1,841	2,093	2,120	2,066	2,410	2,070
Drugs	283	391	501	533	589	623	738	672	739	595	718	1,013
Soaps, cleaners, and toilet goods	284	171	283	387	324	351	331	303	319	329	246	331
Paints and allied products	285	231	175	190	205	293	243	216	228	256	174	163

Table 3 (continued)
ANNUAL QUANTITY OF SOLID WASTES GENERATED IN MANUFACTURING INDUSTRIES[2-12]

Name of industry	SIC Code	1973	1974	1975	1976	1977	1978	1979	1980	1981	1982	1983
Industrial organic chemicals	286	3,894	4,578	6,140	8,376	3,940	3,908	4,015	3,764	5,223	5,448	3,350
Agricultural chemicals	287	2,598	23,385	21,452	30,007	39,177	32,666	28,613	23,203	22,919	19,565	2,777
Miscellaneous chemical products	289	326	508	581	559	519	714	467	530	534	293	391
Petroleum and coal products	29	1,655	2,388	2,038	2,592	2,873	3,585	3,075	4,873	4,667	4,431	3,550
Petroleum refining	291	897	1,472	1,188	1,659	1,715	1,962	1,572	3,800	3,874	3,659	2,997
Paving and roofing materials	295	736	894	831	796	1,131	1,583	1,470	1,041	765	747	508
Rubber and miscellaneous plastic products	30	1,356	1,704	1,404	1,676	1,567	2,056	2,873	2,814	2,926	1,544	1,514
Leather and leather products	31	215	233	221	385	291	290	360	340	295	179	232
Stone, clay, and glass products	32	7,682	9,337	11,345	11,088	12,617	12,742	14,143	13,336	12,098	5,814	6,176
Cement, hydraulic	3,241	3,263	4,579	3,511	3,556	4,318	4,532	4,589	4,568	4,652	2,323	2,482
Concrete, gypsum and plaster products	327	2,154	1,949	3,562	2,756	3,426	3,157	3,126	2,854	2,377	1,318	1,709
Primary metal industries	33	23,973	40,800	42,708	42,353	41,742	46,067	47,783	30,608	35,967	16,670	17,714
Blast furnace and basic steel products	331	16,017	28,045	28,532	25,741	25,141	26,962	28,315	20,180	20,004	9,216	8,379
Iron and steel foundries	332	4,561	6,480	6,149	7,704	8,605	9,987	8,923	7,675	6,768	5,147	6,555
Primary nonferrous metals	333	1,464	5,441	7,180	8,077	7,178	8,208	9,266	1,564	8,325	1,529	1,816
Secondary nonferrous metals	3,341	352	278	288	246	278	246	456	510	251	290	357
Nonferrous rolling and drawing	335	357	396	428	426	400	492	607	440	439	364	414
Nonferrous foundries	336	62	141	100	149	102	122	151	178	137	101	157
Fabricated metal products	34	1,690	2,127	1,933	2,089	1,951	1,992	2,031	1,864	1,801	1,436	1,951
Machinery, except electrical	35	2,253	2,714	2,720	3,067	3,581	3,410	3,524	3,000	2,818	1,630	1,820
Electric and electronic equipment	36	1,505	1,534	1,500	1,500	1,495	1,777	2,319	2,148	1,683	1,250	1,704
Transportation equipment	37	3,818	4,208	3,785	4,283	4,749	5,242	4,253	4,168	3,966	2,893	3,014
Instruments and related products	38	342	552	409	396	418	400	427	480	466	448	569
Miscellaneous manufacturing industries	39	224	319	350	352	335	440	597	527	538	183	296

Note: Values given in thousands of short tons.

Table 4
TOTAL ANNUAL SOLID WASTES GENERATED BY STATE IN THE MANUFACTURING INDUSTRY[2-12]

State	1973	1974	1975	1976	1977	1978	1979	1980	1981	1982	1983
Alabama	691.4	2,854.4	2,702.5	2,851.8	4,042.4	3,281.7	2,801.6	3,495.6	2,875.6	2,254.6	2,715.2
Alaska	25.8	57.5	77.0	86.3	98.5	90.8	107.6	85.6	91.7	66.9	2,715.2
Arizona	1,110.9	4,921.3	3,881.1	5,305.4	4,936.0	6,121.7	6,936.0	616.2	7,107.4	133.8	452.0
Arkansas	596.8	2,025.8	1,940.8	2,105.0	2,390.1	2,520.8	2,636.4	2,522.7	2,458.4	1,219.9	242.6
California	5,966.0	5,961.0	7,272.0	6,191.0	7,968.0	9,547.0	8,106.0	7,811.0	7,097.0	3,493.0	5,080.4
Colorado	899.2	607.9	866.9	1,149.6	1,306.8	1,269.3	1,343.9	1,178.2	1,524.1	556.2	545.9
Connecticut	513.6	606.9	528.5	559.1	584.0	542.0	578.6	618.9	560.3	556.9	946.1
Delaware	212.6	239.3	221.5	329.2	250.8	311.5	322.3	339.7	—	297.0	746.4
Florida	2,390.2	21,303.6	14,587.2	23,465.2	27,583.6	20,968.6	14,996.0	13,855.1	13,722.8	13,516.8	1,247.5
Georgia	825.1	1,686.1	1,817.1	1,872.3	1,734.5	1,991.3	1,642.7	1,723.9	1,720.3	1,599.5	1,996.1
Hawaii	1,542.4	2,017.5	2,467.2	3,688.5	2,504.5	2,625.2	2,431.8	1,882.4	2,681.7	1,770.6	2,004.7
Idaho	1,621.5	2,858.1	3,947.4	3,460.3	4,382.0	3,891.4	4,201.3	1,626.2	3,235.0	3,332.1	1,256.5
Illinois	4,974.9	5,258.1	5,191.8	4,681.4	5,473.7	6,831.7	6,330.9	6,150.8	5,914.2	3,272.2	3,420.5
Indiana	4,021.1	5,397.4	4,980.6	9,237.8	7,920.3	9,783.6	8,168.6	5,426.3	4,377.2	3,551.5	3,290.6
Iowa	990.4	1,725.2	6,023.2	1,946.1	1,669.4	1,723.9	1,983.0	1,137.1	1,170.1	599.4	886.7
Kansas	258.1	711.6	756.7	836.6	741.4	789.2	—	1,358.3	796.1	859.9	403.1
Kentucky	1,014.3	998.7	1,203.1	1,188.0	1,292.6	1,588.7	1,564.6	1,719.7	1,656.5	805.5	1,190.4
Louisiana	1,552.8	2,962.4	3,537.9	8,160.4	9,489.4	9,250.4	9,206.3	9,938.1	8,927.0	6,832.8	4,029.7
Maine	101.9	772.4	567.0	2,911.9	862.0	1,018.0	1,689.8	989.7	1,125.0	1,045.5	1,417.9
Maryland	628.3	612.8	603.4	754.2	1,222.0	1,914.7	2,024.1	1,966.4	1,324.7	1,338.2	1,030.1
Massachusetts	524.3	675.6	704.3	768.5	923.9	1,402.4	1,394.9	1,340.7	1,266.9	663.8	1,148.7
Michigan	5,936.9	6,808.3	6,453.2	8,115.7	9,074.3	8,638.5	8,370.9	7,403.9	7,181.4	4,928.6	4,611.3
Minnesota	1,019.5	1,115.7	1,014.8	1,027.2	877.8	1,052.6	888.7	924.6	969.3	895.8	986.9
Mississippi	1,513.2	1,695.6	1,966.9	2,358.4	2,147.4	2,183.9	2,026.1	2,092.8	2,531.5	1,273.6	1,379.2
Missouri	1,199.0	2,037.5	1,906.2	1,192.4	1,241.4	1,133.9	3,829.6	3,343.4	2,286.2	1,408.4	1,471.4
Montana	367.9	931.2	360.9	418.7	383.1	358.1	486.8	546.5	632.2	388.6	444.6
Nebraska	289.1	598.7	629.0	560.5	640.4	683.7	815.3	865.8	856.9	231.6	391.9
Nevada	148.0	298.2	128.2	267.0	30.0	—	—	—	35.1	28.2	25.8
New Hampshire	175.3	204.2	199.5	323.5	412.9	—	—	—	—	192.1	237.3
New Jersey	1,464.7	1,909.0	1,813.5	1,914.2	2,027.7	2,391.6	4,823.5	3,608.9	2,821.5	2,034.8	1,905.5
New Mexico	20.4	80.3	—	42.7	17.1	—	—	1,619.5	1,701.2	—	17.7
New York	2,902.9	4,434.4	4,422.6	4,506.0	4,369.7	4,276.1	5,841.9	4,710.9	4,575.1	2,501.8	4,115.9

Table 4 (continued)
TOTAL ANNUAL SOLID WASTES GENERATED BY STATE IN THE MANUFACTURING INDUSTRY[2-12]

State	1973	1974	1975	1976	1977	1978	1979	1980	1981	1982	1983
North Carolina	1,382.1	1,541.9	1,647.7	1,620.6	2,165.7	2,674.2	2,583.2	2,365.2	1,741.0	2,460.7	2,511.1
North Dakota	155.4	231.4	176.4	228.0	44.6	—	—	—	24.5	18.4	40.9
Ohio	9,528.1	9,042.7	9,107.8	8,503.1	7,076.1	7,910.5	7,856.0	6,849.3	6,682.7	5,761.6	6,862.8
Oklahoma	236.7	434.2	447.2	675.8	408.2	464.6	—	525.2	657.8	555.8	1,641.0
Oregon	1,002.2	1,458.8	2,055.9	2,434.4	1,988.7	2,045.4	2,233.0	1,450.3	1,317.6	1,382.6	1,748.4
Pennsylvania	4,259.5	11,900.7	11,660.1	12,569.1	11,655.2	11,365.1	11,324.6	8,606.3	8758.1	5,198.9	5,151.2
Rhode Island	83.6	104.9	115.9	150.9	98.4	105.6	133.3	75.2	92.3	77.3	72.5
South Carolina	493.2	688.4	737.0	874.8	1,036.6	1,128.2	1,445.0	1,131.3	1,383.1	990.6	1,231.3
South Dakota	17.3	21.0	24.2	28.3	103.1	—	—	—	18.6	19.0	152.4
Tennessee	1,596.0	4,667.4	7,892.1	2,809.4	3,492.6	4,174.7	4,583.8	5,091.6	4,304.4	2,630.1	1,607.6
Texas	5,864.4	8,139.4	9,428.4	11,087.0	10,552.3	9,368.6	13,345.1	13,313.5	14,336.2	9,464.1	8,878.8
Utah	73.8	1,105.9	1,353.9	823.7	1,216.8	1,077.1	952.3	726.6	796.0	182.7	103.1
Vermont	54.7	50.5	39.3	43.8	48.0	—	—	—	—	31.7	26.9
Virginia	1,520.8	1,720.7	1,856.6	1,799.3	1,461.1	1,532.6	1,728.6	1,620.5	1,783.6	1,986.6	2,010.5
Washington	1,994.7	1,869.8	3,860.8	4,882.2	3,553.3	2,982.8	3,237.0	3,272.2	2,904.2	2,068.5	2,010.9
West Virginia	2,764.1	3,391.5	2,797.2	2,888.9	2,861.4	2,978.6	3,117.2	2,697.5	3,220.8	2,391.7	1,318.9
Wisconsin	1,695.2	2,392.4	2,721.3	2,929.3	3,225.6	3,632.4	—	3,461.0	3,635.1	2,393.6	2,692.5
Wyoming	246.4	182.0	—	—	—	—	—	—	230.9	146.0	114.1

Note: Values given in thousands of short tons.

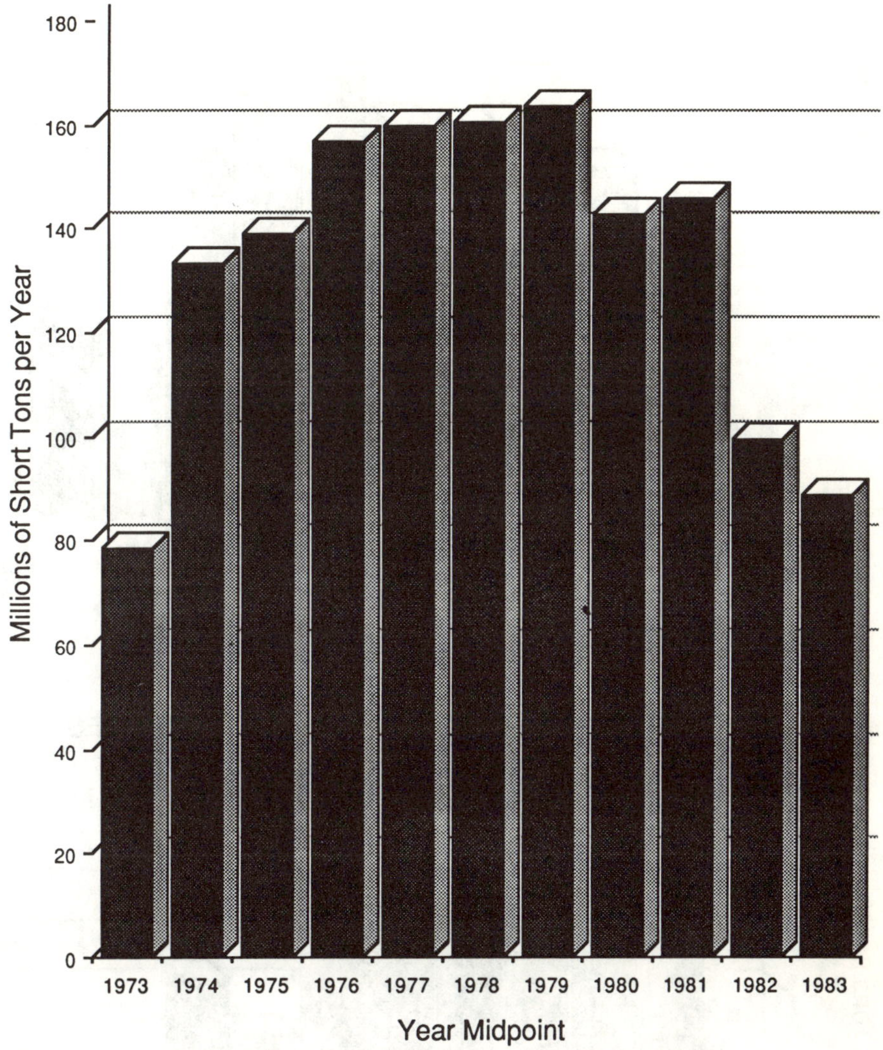

FIGURE 1. U.S. total annual solid waste generation by all manufacturing industries.[2-12]

The meat products subgroup of the industry includes processing plants that slaughter livestock and other animals and then process the meat as either fresh, frozen, or canned consumer products. The dairy products subgroup includes processing plants that produce milk products and prepare milk for distribution. Fruits, vegetables, and seafoods are processed as bottled products, dried foods, frozen or canned items. Also included are grain mills for producing flour, establishments that produce cereals and baked goods, plants that produce raw or finished sugar from sugar cane and sugar beets, and plants that produce confectionaries and beverages.

B. Manufacturing Processes

The food and kindred products manufacturing industry uses natural materials to extract and produce refined food items. The processes involved range from mechanical operations to alternately cooking, mixing, and straining to create the products. Precleaning and removing undesirable materials from foods prior to processing is almost always needed. This precleaning is commonly done by washing with water. Following

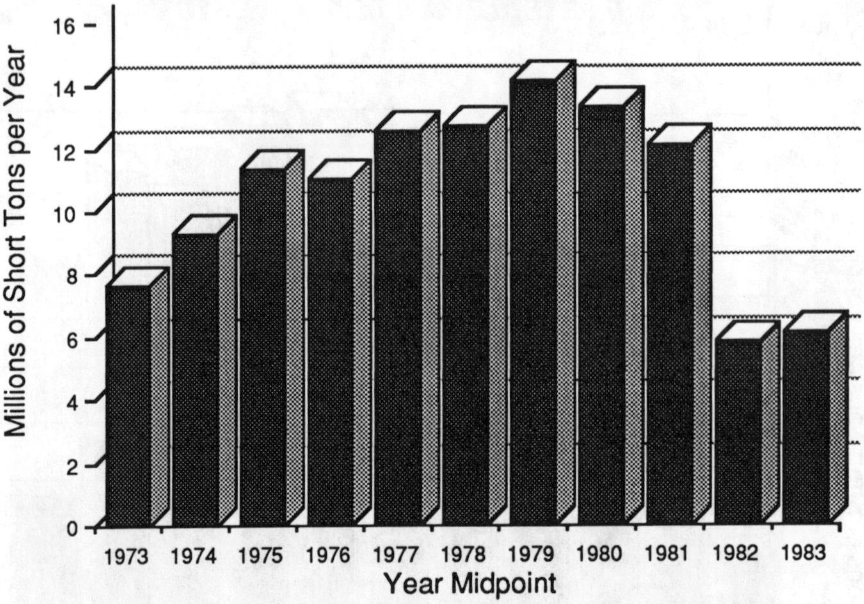

A

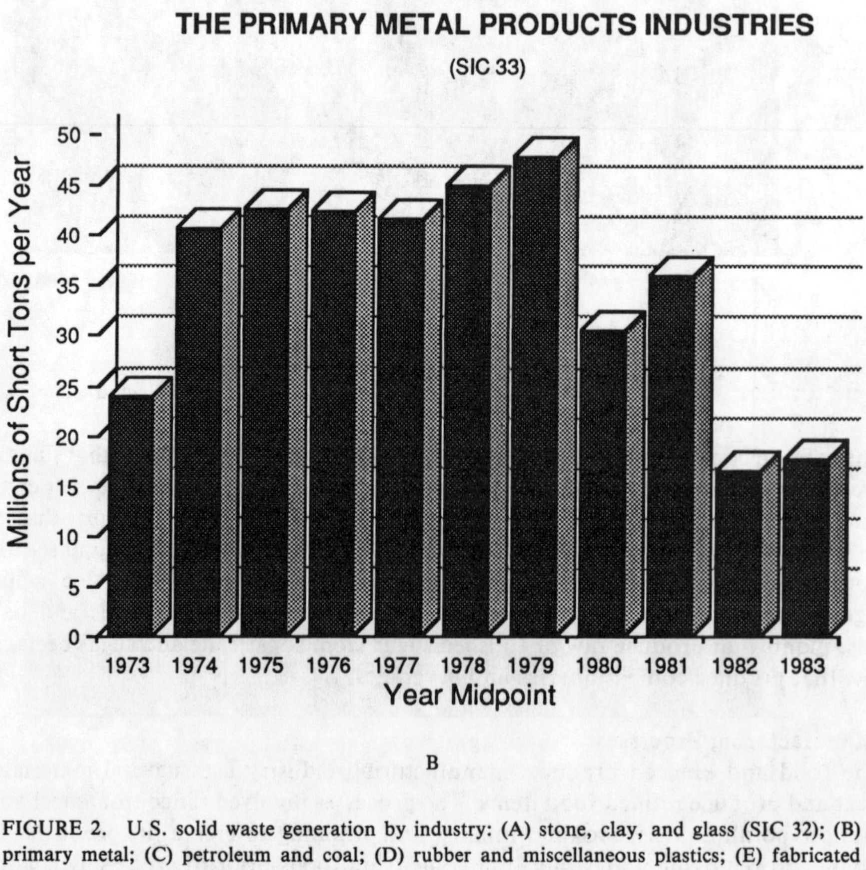

B

FIGURE 2. U.S. solid waste generation by industry: (A) stone, clay, and glass (SIC 32); (B) primary metal; (C) petroleum and coal; (D) rubber and miscellaneous plastics; (E) fabricated metal; (F) paper; (G) printing and publishing.[2-12]

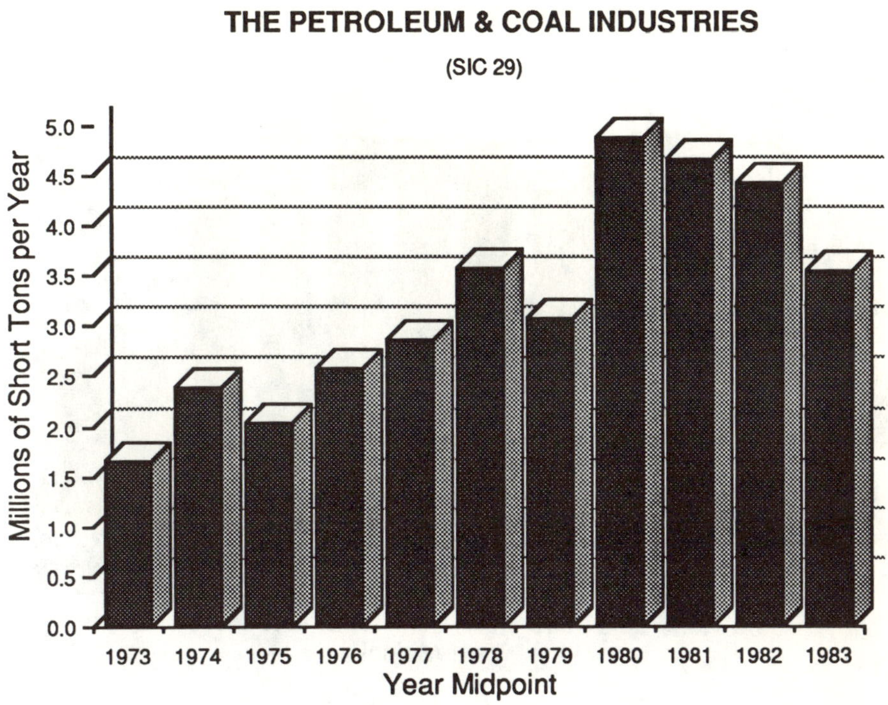

FIGURE 2C.

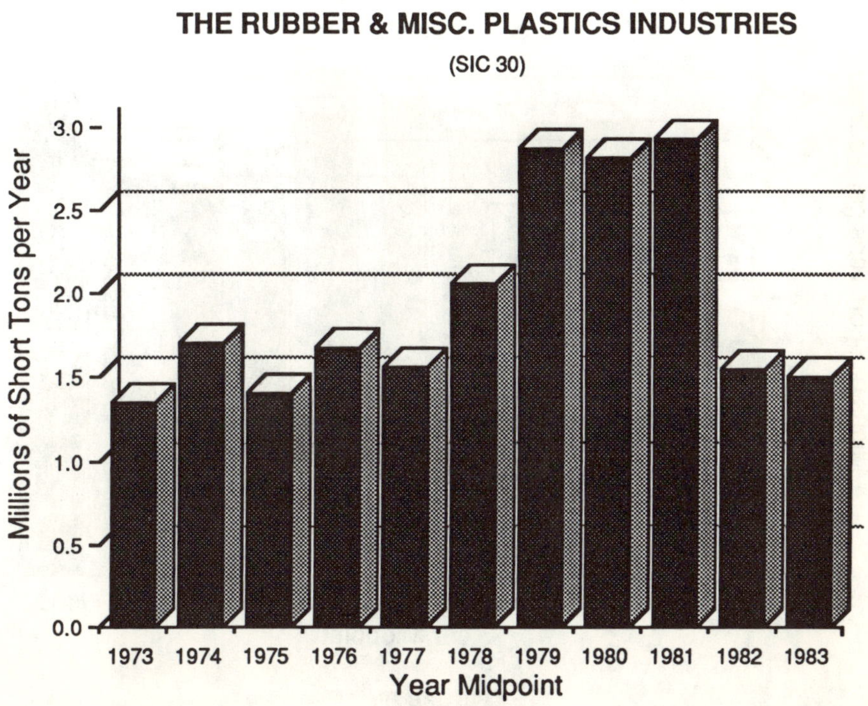

FIGURE 2D.

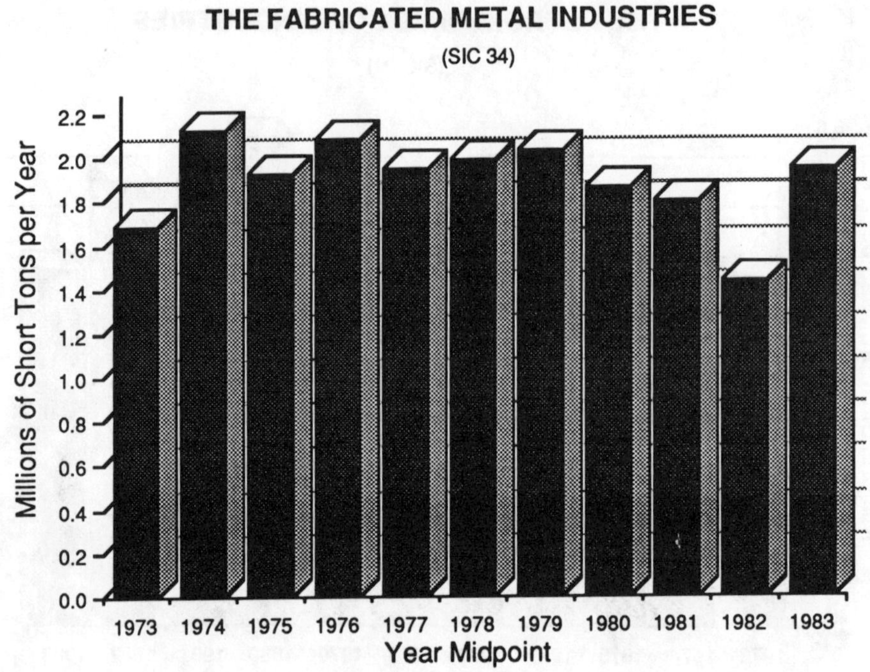

FIGURE 2E.

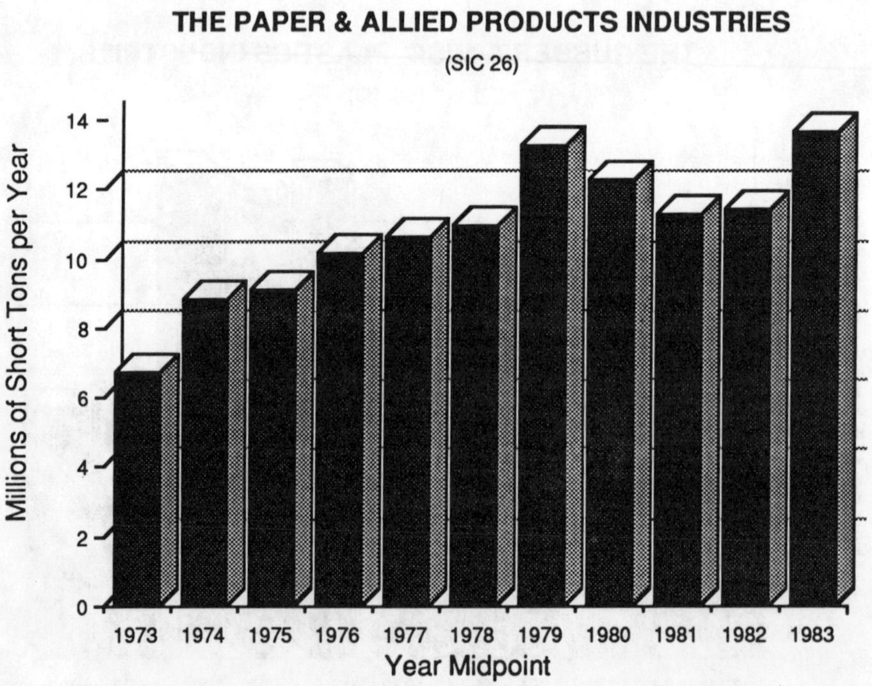

FIGURE 2F.

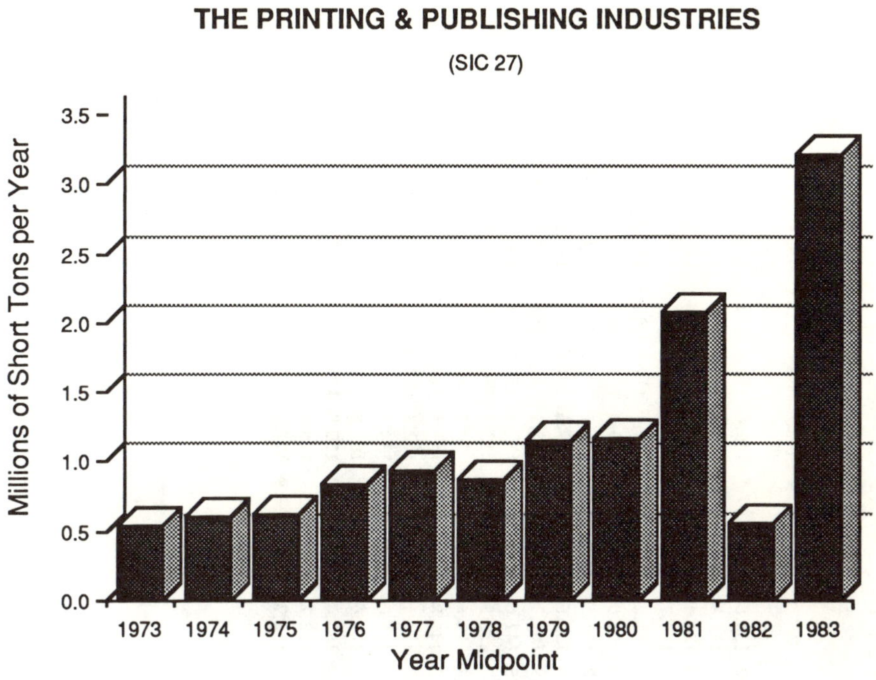

FIGURE 2G.

the washing step, most food materials require trimming, peeling, sizing, cutting, and coring. Wastes are produced during each of these steps. The wastewater usually contains large amounts of organic matter and suspended solids. The treatment of these wastewaters yields organic sludge. As an example of a food and kindred product industry, dairy industry manufacturing and disposal practices are described below.

The dairy products subgroup (SIC 202) processes raw milk into market milk, butter, cheese, condensed and evaporated milk, and frozen desserts. Market milk refers to milk which has been pasteurized, homogenized, and packaged for sale to households and commercial establishments. The first step in the dairy industry is delivering the milk to the dairy plant where it is then transferred to refrigerated storage tanks. The next step is to pasteurize the milk. This is a heat treatment designed to kill pathogenic microorganisms. It has traditionally been done by heating milk to 63°C for about 30 min. Modern plants heat milk to 72 to 75°C for 15 seconds or 135 to 150°C for 3 to 5 seconds.[16] This is followed by clarification and separation. A centrifuge is used to remove particles and separate this skim milk and the cream. The milk fat concentration is standardized by further separation of the cream and skim milk from the whole milk. The milk is then homogenized to reduce the fat particles to a small enough size so that they remain dispersed throughout the liquid. Finally, volatile substances are removed by deaeration.

Processing of the raw milk into the various dairy products involves a combination of these and additional procedures; e.g., in buttermaking after separation and pasteurization of the cream, the milk fat is allowed to crystallize. The crystallized cream is then churned into butter. It is then mixed, blended, and packaged for sale.

C. Waste Disposal and Reuse Practices

In general, the solid wastes generated by the food and kindred product industries are either buried in landfills, placed in lagoons, sprayed on land, incinerated, composted

Table 5
U.S. ANNUAL MANUFACTURING PRODUCTION BY INDUSTRY GROUP[14,15]

SIC Code	Category	1967	1972	1973	1974	1975	1976	1977	1978	1979	1980	1981	1982	1983	1984
20	Foods	100.0	116.9	120.9	124.0	123.4	133.0	138.0	142.7	147.5	149.6	152.1	151.1	156.4	163.2
21	Tobacco products	100.0	105.7	112.3	109.8	111.8	116.9	112.8	118.3	117.8	119.9	122.2	118.0	112.1	115.2
22	Textile mill products	100.0	132.7	142.9	132.8	122.3	134.6	134.4	137.5	145.0	138.6	135.7	124.5	140.8	138.6
23	Apparel products	100.0	109.4	117.3	114.3	107.6	125.7	125.7	134.2	134.4	127.0	120.4	—	—	—
24	Lumber and products	100.0	120.8	126.0	116.2	107.6	123.2	131.2	136.3	136.9	119.3	119.1	112.6	137.2	148.7
25	Furniture and fixtures	100.0	130.6	143.7	137.6	118.2	136.8	145.0	155.8	161.5	150.0	157.2	151.9	170.5	190.2
26	Paper products	100.0	128.8	137.4	134.5	116.3	133.2	137.5	144.8	151.0	151.1	155.0	150.8	164.3	174.4
27	Printing and publishing	100.0	112.7	118.2	118.2	113.4	122.5	127.6	131.5	136.9	139.6	144.2	144.1	152.5	169.7
28	Chemical and products	100.0	143.6	154.5	159.4	147.3	170.9	185.7	197.4	211.8	207.1	215.6	196.1	215.0	228.1
29	Petroleum products	100.0	122.1	128.5	124.7	124.1	133.5	142.6	145.2	143.9	132.9	129.7	121.8	120.3	124.4
30	Rubber and plastic products	100.0	172.4	184.1	195.2	166.7	199.7	232.3	253.6	272.2	255.7	274.0	254.7	291.9	331.7
31	Leather and products	100.0	88.1	83.1	78.1	76.5	77.4	73.6	73.8	71.7	70.1	69.3	60.9	61.9	59.9
32	Clay, glass, and stone products	100.0	120.8	133.6	133.1	117.9	136.2	145.8	157.2	163.9	147.5	147.9	128.2	143.4	159.7
33	Primary metal products	100.0	112.1	126.7	123.1	96.4	109.7	111.1	119.9	121.3	102.3	107.9	75.3	85.4	95.1
34	Fabricated metal products	100.0	112.1	124.7	124.2	109.9	123.9	131.0	141.6	148.5	134.1	136.4	114.8	120.2	137.5
35	Nonelectrical machinery	100.0	116.0	133.7	140.1	125.1	134.5	143.6	153.6	163.7	162.8	171.2	149.0	150.6	181.5
36	Electrical machinery	100.0	122.3	143.1	143.8	116.5	134.8	145.4	159.4	175.0	172.8	178.4	169.3	185.5	217.4
37	Transportation equipment total	100.0	108.2	118.3	108.8	97.4	111.1	122.2	132.5	135.4	116.9	116.1	104.9	117.8	137.6
38	Instrument	100.0	120.4	137.7	144.1	132.3	146.6	156.2	167.1	174.9	171.1	170.3	161.9	158.7	174.2

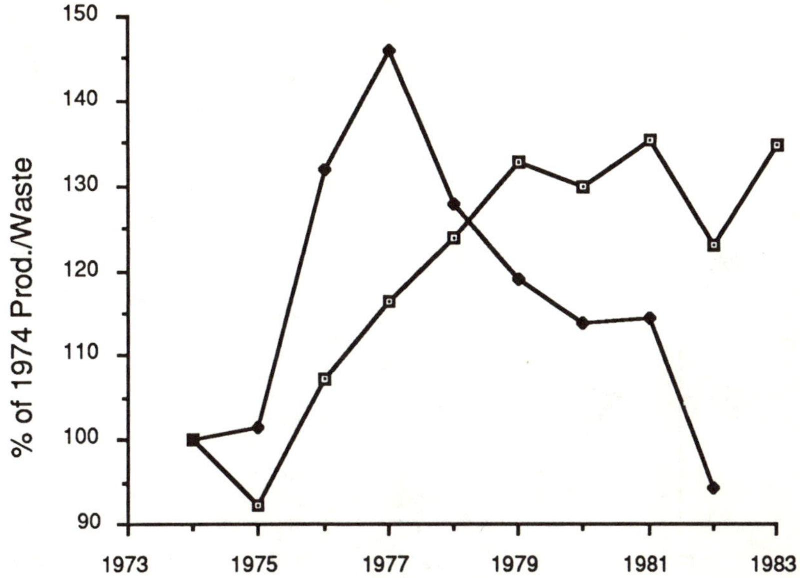

A

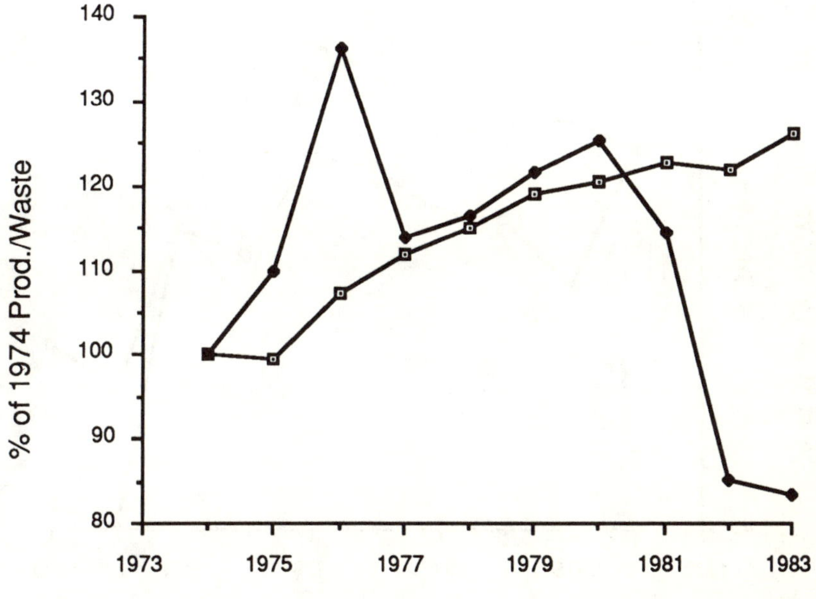

B

FIGURE 3. Normalized annual production and solid waste generation (a) chemicals, (b) food, (c) paper and (d) primary metals. ⊡, Normalized annual production; ■, normalized solid waste generation.[2-15]

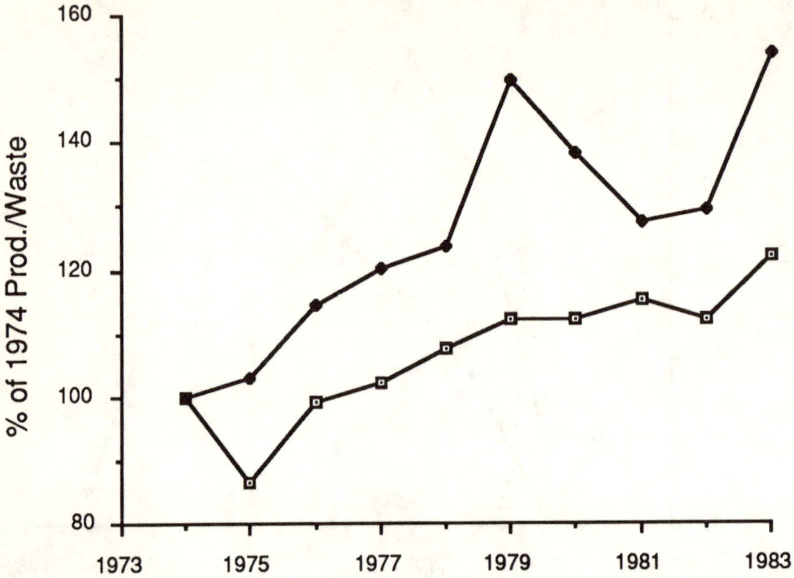

FIGURE 3C.

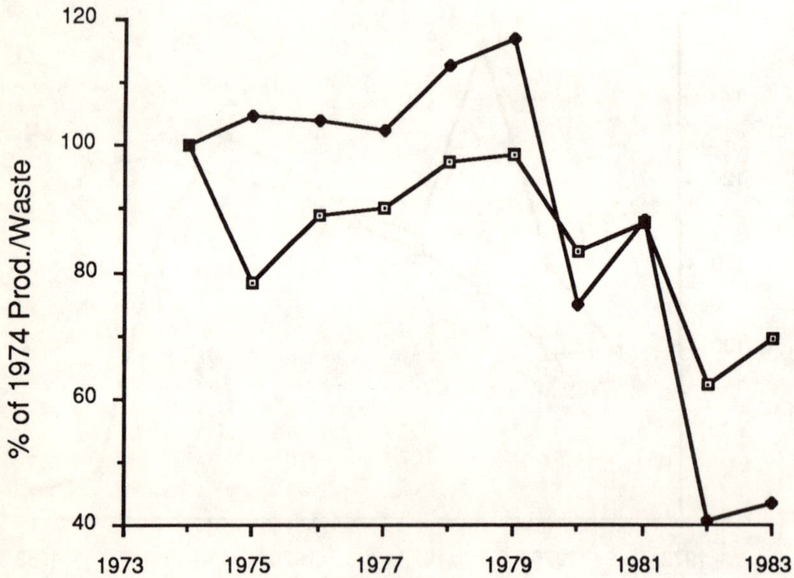

FIGURE 3D.

to create a humuslike material, or discharged to surface waters after grinding and liquifying. Most of these solid wastes are placed in municipal sanitary landfills. Since data on the chemical composition of the solid waste could not be obtained, a general description of chemicals which should be expected in the various waste components are given in Table 7. This list includes pesticide and fertilizer residuals as well as food residuals.[17]

The main ingredient in the dairy industry is milk. The wastes produced are typically dilutions of milk and milk products. The dairy industry generates large volumes of

Table 6
ANNUAL QUANTITY OF SOLID WASTES GENERATED BY STATE IN THE FOOD AND KINDRED PRODUCTS MANUFACTURING INDUSTRY[2-12]

State	1973	1974	1975	1976	1977	1978	1979	1980	1981	1982	1983
Alabama	29.9	36.4	26.0	161.6	154.7	—	—	—	—	77.3	69.7
Alaska	0.1	1.0	1.0	6.6	30.0	—	—	—	—	—	69.7
Arizona	19.1	7.5	19.8	15.9	13.1	—	—	—	—	—	82.7
Arkansas	80.8	67.4	90.7	126.5	59.4	80.4	—	69.4	93.1	57.3	10.3
California	1482.0	1564.0	2039.0	1889.0	1920.0	2268.0	1823.0	2071.0	2026.0	1413.0	1193.5
Colorado	210.9	237.9	228.3	193.3	134.4	112.1	—	101.0	175.5	—	231.7
Connecticut	46.5	20.1	13.2	15.8	16.4	—	—	—	—	—	0.0
Delaware	18.3	16.6	21.7	31.8	16.6	—	—	—	—	32.8	23.9
Florida	367.0	878.9	564.1	614.8	676.1	661.2	379.0	489.7	467.8	302.6	347.1
Georgia	82.0	110.7	122.2	205.7	137.0	106.7	262.6	83.7	94.9	88.9	148.0
Hawaii	1536.1	2015.4	2405.7	—	—	—	—	1813.2	—	—	1999.5
Idaho	216.4	445.4	327.9	532.9	523.1	—	—	566.8	—	—	88.6
Illinois	370.5	483.5	449.2	540.5	515.8	780.2	677.2	918.4	870.6	365.6	470.9
Indiana	124.1	159.4	165.2	242.3	116.0	126.0	111.1	134.5	127.9	125.1	131.4
Iowa	295.2	365.4	1047.7	1086.5	358.2	321.1	582.3	385.6	457.3	—	222.5
Kansas	57.5	68.1	77.9	79.2	89.2	—	—	113.6	100.4	27.0	34.0
Kentucky	31.4	46.3	50.8	59.8	90.4	48.1	—	41.4	73.5	50.1	51.1
Louisiana	—	152.6	231.4	308.4	279.0	327.1	257.6	168.5	197.2	182.8	202.2
Maine	38.0	31.7	90.8	32.4	10.7	—	—	—	—	10.4	23.6
Maryland	37.0	46.1	73.2	45.1	52.5	60.9	61.1	—	—	56.1	50.4
Massachusetts	77.5	86.4	82.7	77.0	52.8	—	46.3	—	—	45.8	76.7
Michigan	370.5	251.5	150.8	243.9	248.5	281.1	343.3	371.0	—	487.4	339.6
Minnesota	322.9	256.4	241.6	183.9	172.4	256.3	163.4	202.7	308.3	253.3	249.4
Mississippi	23.7	29.1	44.1	38.6	32.0	—	—	—	—	162.6	185.1
Missouri	193.6	212.4	173.3	153.7	160.7	164.6	122.4	128.7	120.2	100.4	237.9
Nebraska	136.0	432.8	376.1	285.3	509.0	532.8	637.3	666.9	619.5	72.4	221.9
Nevada	—	1.9	—	—	0.8	—	—	—	—	—	0.0
New Jersey	175.1	291.3	223.9	155.6	217.5	253.6	195.4	218.5	223.1	315.9	93.1
New York	245.5	285.4	288.7	207.6	332.5	356.4	436.9	249.2	266.9	289.4	428.6
North Carolina	30.6	40.2	33.5	36.4	51.6	47.9	36.3	43.8	—	65.2	75.4
Ohio	264.9	286.8	265.1	247.1	416.6	317.5	300.1	297.4	554.5	245.0	292.9

Table 6 (continued)
ANNUAL QUANTITY OF SOLID WASTES GENERATED BY STATE IN THE FOOD AND KINDRED PRODUCTS MANUFACTURING INDUSTRY[2-12]

State	1973	1974	1975	1976	1977	1978	1979	1980	1981	1982	1983
Oklahoma	13.1	24.1	14.4	14.9	40.2	—	—	—	—	16.6	48.4
Oregon	120.3	214.4	198.9	418.0	240.7	—	—	—	81.7	341.1	129.3
Pennsylvania	217.3	318.8	308.3	359.4	294.9	337.0	326.5	295.4	249.6	353.9	293.3
South Carolina	6.4	8.0	17.1	25.2	22.9	—	—	—	—	15.2	13.3
South Dakota	—	14.9	15.5	18.2	92.5	—	—	—	—	—	0.0
Tennessee	89.8	113.8	115.9	101.3	116.3	104.6	125.3	129.5	104.2	100.1	74.1
Texas	260.0	236.8	202.0	350.1	1002.5	588.8	1299.6	—	502.0	531.0	312.7
Utah	41.5	47.1	—	44.3	59.7	—	—	—	—	—	0.0
Vermont	—	—	4.6	—	—	—	—	—	—	0.7	0.0
Virginia	148.7	104.4	129.6	123.1	79.0	—	—	76.9	67.7	59.4	76.2
Washington	419.5	615.6	683.8	961.5	508.6	531.6	418.6	423.4	443.3	269.9	208.0
West Virginia	4.2	3.0	14.7	18.0	24.6	—	—	—	—	1.8	1.8
Wisconsin	373.9	370.6	508.1	534.2	500.4	462.8	492.2	—	534.5	332.8	288.3
Wyoming	—	170.4	181.8	—	—	—	—	—	—	—	0.0

Note: Values given in thousands of short tons.

Table 7
QUALITATIVE CHEMICAL COMPOSITION OF SOLID WASTES FROM THE FOOD AND KINDRED PRODUCTS INDUSTRY[17]

Source waste	Chemical content
Fruit and vegetable remains Grain and fruit residues in brewery winery	Aldehydes, ketones, organic acids, phenol, nitrogen, sulfur, phosphorous
Meat residues Fish residues Milk residues	Aldehydes, ketones, organic acids, phenols, sulfates, phosphate, carbonate, nitrites, nitrates, protein, carbohydrate, fatty acids
Beet and sugar cane residues	Aldehydes, ketones, organic acids, phenol, nitrogen, sulfur, phosphorous and residual pesticides, carbohydrates, proteins, lignins and trace elements

wastewater and sludges. In the U.S., for each volume of dairy products up to seven volumes of waste-water are generated.[16] There are three types of sludges: separator sludge, primary treatment sludge, and secondary treatment sludge.

The separator sludge is produced in the centrifuges used to separate the cream from the milk. About 0.01 and 0.1% of the milk entering the separator becomes separator sludge. Separator sludge is typically heat treated to kill pathogenic organisms and then landfilled, used as livestock feed, or incinerated.

The primary treatment of wastewaters produces sludges from settling of suspended solids. Typically, the solids are settled by chemical flocculation. This produces a large volume of sludge containing fat and protein.[16] These sludges are normally disposed of in landfills.

Further treatment of dairy wastewaters (secondary treatment) using biochemical treatment processes results in more sludge. The sludge is then sprayed on land, disposed of in landfills, or incinerated. Dairy wastes have also been used as fertilizer.[16]

IV. TEXTILE MILL PRODUCTS (SIC CODE 22)

A. Waste Generation

This industry processes wool, cotton, and synthetic fibers through weaving and finishing. Table 8 summarizes solid waste generation for the textile industry by state and by year.[2,-12] Solid waste generation in the textile industry peaked in 1978. This industry produces less than 1% of the total solid wastes generated by all manufacturing industries in 1983. The textile industry is concentrated in the southern U.S., with the largest quantities of wastes generated in North Carolina, South Carolina, and Georgia.

B. Manufacturing Processes

1. Wool Processing Industry

Wool industry processing (Figure 4) begins with scouring or washing of raw sheepswool. The washed wool is then carbonized to remove additional wastes before a final washing and neutralization with sodium carbonate. The wool fibers subsequently undergo carding, fulling, and weaving. Dying can be done before or after the weaving step. Dyes used are either acid dyes, mardant dyes or metallic dyes. During the scouring of wool fibers, solvents (e.g., methyl alcohol, isopropyl, or ethyl alcohol) or detergents (e.g., sodium carbonate, nonionic detergents of ethylene oxide class) are used to remove grease and remaining dirt. During the carbonizing step, mineral acids are used to degrade cellulose and hydrocellulose in the wool. Except for grease, most of the

Table 8
ANNUAL QUANTITY OF SOLID WASTES GENERATED BY STATE IN THE TEXTILE MILL PRODUCTS MANUFACTURING INDUSTRY[2-12]

State	1973	1974	1975	1976	1977	1978	1979	1980	1981	1982	1983
Alabama	30.0	35.2	26.5	18.2	22.7	—	—	—	—	52.8	58.4
Arkansas	1.7	1.4	1.8	—	—	—	—	—	—	8.1	0.0
California	5.0	6.0	6.0	6.0	37.0	—	—	—	—	—	9.8
Connecticut	6.9	18.6	16.8	14.2	13.7	—	—	—	—	—	18.4
Georgia	64.9	77.2	93.9	96.9	94.2	—	103.5	90.8	95.3	63.6	99.5
Maine	2.2	1.4	1.2	2.8	1.7	—	—	—	—	—	7.5
Massachusetts	18.2	26.5	27.4	35.1	24.3	—	—	—	—	9.7	25.5
Minnesota	3.7	0.5	0.7	0.7	1.3	—	—	—	—	—	0.0
Missouri	4.0	3.6	2.7	0.6	—	—	—	—	—	—	0.0
New Jersey	13.0	18.6	31.5	16.6	25.0	—	—	—	—	9.9	23.0
New York	17.8	24.0	11.7	11.8	59.6	—	—	—	—	—	3.2
North Carolina	130.2	174.3	155.9	128.5	236.3	408.1	414.6	228.4	219.8	89.1	116.4
Pennsylvania	32.3	76.6	53.3	44.7	22.3	—	22.4	—	—	60.7	43.2
Rhode Island	5.1	2.3	13.0	9.5	8.4	—	—	—	—	—	24.5
South Carolina	95.2	160.0	111.5	131.1	109.4	163.0	170.4	121.2	99.1	74.1	101.0
Tennessee	15.8	21.3	19.3	14.4	19.9	—	17.3	—	—	—	6.5
Virginia	28.6	20.5	31.8	59.2	80.6	71.4	70.0	72.3	65.4	35.2	33.4
Wisconsin	2.0	1.9	—	—	—	—	—	—	—	—	0.0

Note: Values given in thousands of short tons.

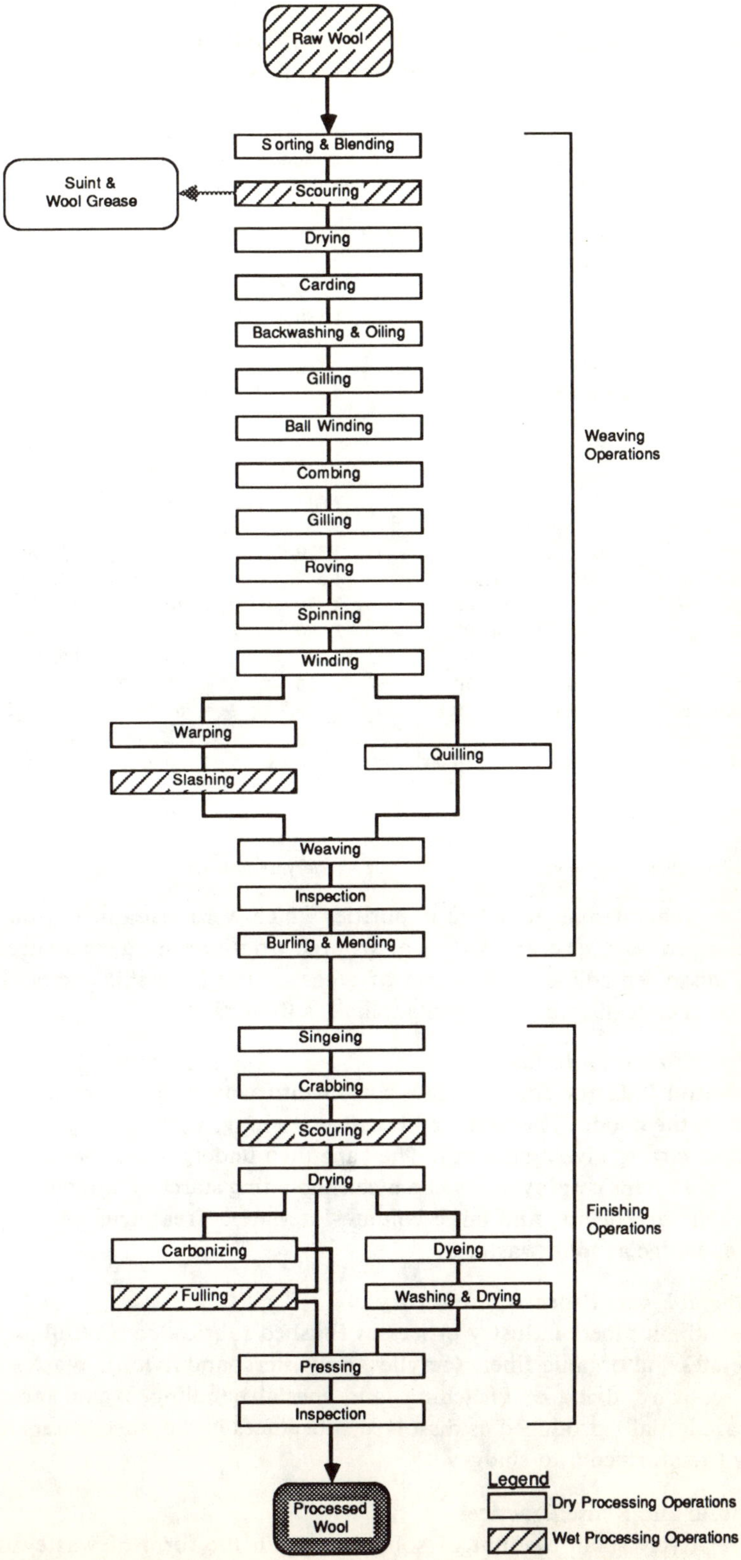

FIGURE 4. Schematic diagram of wool processing.

Table 9
TYPICAL CONCENTRATIONS OF CHEMICALS FOUND IN TEXTILE MILL SOLID WASTES[19]

Parameter	Wool scouring	Wool fabric dyeing and finishing	Woven fabric dyeing and finishing	Knit fabric dyeing and finishing	Carpet dyeing and finishing	Yarn and dyeing and finishing
Arsenic	<0.1	<17	<1	<4.8	<10	<2.5
Barium	59	<170	39	<53	<95	25
Cadmium	1.2	<17	4.4	4.5	<10	<2.5
Chromium	19	267	1,196	33	112	31
Cobalt	4.2	<67	26	<23	124	12
Copper	18	117	652	410	211	264
Iron	4,820	1,100	4,910	3,840	5,200	1,660
Lead	28	<170	36	<52	110	<25
Manganese	205	8,000	128	51	256	66
Mercury	<0.01	<1.7	0.35	1.4	<198	0.66
Molybdenum	<2	<333	<17	<94	—	50
Nickel	12.5	<33	32	<25	<49	12
Zinc	106	1,130	2,370	550	1,790	1,505
Aluminum	4,860	11,500	4,640	3,180	4,430	1,320
Magnesium	5,560	12,000	2,820	1,210	1,820	3,590
Potassium	9,240	14,000	3,580	2,850	4,020	4,770
Sodium	675	137,000	51,300	54,200	66,100	359,000
Strontium	21.6	170	16	15	33	40
Total chlorinated organics	1.26	0.11	15.2	64.7	26.2	40.1
Suspended solids (%)	9.8	0.008	0.88	0.69	0.024	0.015
Total solids (%)	10.1	0.06	1.26	0.87	0.11	0.165

Note: Sludge values given in ppm (dry weight) unless otherwise noted.

wastes are generated as dissolved impurities which were contained in the raw wools.[18] It is estimated that for every 450 kg of scoured wool produced, there are about 225 kg of total dissolved solids in the form of grease, suint (oily skin secretions), dirt, and processed chemicals (e.g., detergents, alkali, softeners).[17]

2. Cotton Processing Industry

The cotton industry converts raw cotton into woven cloth or yarn and then into finished textile goods. The first step is called slashing, where sizing material is applied to the raw yarn to give it strength. The yarn then undergoes scouring and dying operations. These steps employ many chemicals including starches, polyvinyl alcohol, acids and alkaline detergents, and large volumes of water. Treatment of these wastewaters generates sludge as solid waste.

3. Synthetic Fibers Processing Industry

The synthetic fibers industry processes finished fabrics from cellulose fibers (rayon and acetate) and organic fibers (acrylics, polyesters, and nylon). Wastes are generated during scouring, dying or bleaching, and special handling. Again most of the solid wastes are initially produced as dissolved substances in the wastewaters and are subsequently transformed into sludges.

C. Disposal and Reuse Practices

The relatively small amount of solid wastes, in the form of wastewater treatment sludge generated by the textile mill products group is generally disposed of in landfills through municipal systems. Chemical analysis for these sludges is included in Table 9.

Nickel, zinc, aluminum, magnesium, and potassium concentrations are high.[19] Wool grease and suint wastes can be reused to recover lanolin and potash; however, such reuse is very sensitive to economic considerations.

V. LUMBER AND WOOD PRODUCTS (SIC 24)

The lumber and wood products industry includes saw mills and planing mills (SIC 242); millwork, plywood and structural members (SIC 243); and wood containers (SIC 244). As examples, both wood preservation (SIC 2491) and hardboard manufacturing (SIC 2499) are discussed.

A. Waste Generation

Solid waste generation in the lumber and wood products industry has averaged 6.2 million tons per year. Solid waste generation peaked in 1976 at 9.3 million tons and remained around 6 million tons between 1977 and 1981.[2-12] In 1982 and 1983, solid waste production decreased to about 3.7 million and 4 million tons, respectively. In 1983, the lumber and wood products industry generated about 4.5% of the total manufacturing of solid wastes. Table 10 lists solid wastes generated by state and year for this industry group.[2-12]

B. Manufacturing Processes

1. Wood Preservation (SIC 2491)

The use of wood preservatives increases the useful life of the wood by injecting chemicals which act as fungicides, insecticides and fire retardants. Preservatives are added to woods which are used outdoors such as telephone poles, railroad ties, and fence posts. The most common chemicals used are copper, chromium and arsenic salts, creosote, and pentachlorophenol. Borates, phosphates, and ammonium compounds are sometimes used as fire retardants.[19]

Wood preservation is completed in two basic steps: (1) preconditioning the wood and (2) impregnating the wood. The purpose of preconditioning is to reduce the moisture content and increase the void volume of the wood. Preconditioning methods include drying the wood in large open yards, kiln drying, and steam conditioning. Following preconditioning, the wood is treated by either nonpressure or pressure preserving processes. Nonpressure treatment involves soaking the wood in hot or cold vats of the preservative. Pressure treatment processes use a combination of air and hydrostatic pressure.[20]

2. Manufacturing Hardboard (SIC 2499)

Hardboard is a type of fiberboard composed of ligno-cellulosic fibers. It is used for paneling, siding, and furniture. There are two major hardboard manufacturing processes, the wet process and the dry process. This section discusses the wet process because it produces a large quantity of wastewater and, consequently, large volumes of wastewater treatment sludges.

Hardboard is primarily composed of wood and wood scraps. The manufacturing process is summarized in Figure 5. First the wood is converted into chips. The chips are steamed in the preheater to soften the wood. Next, the chips are prepared in a thermal mechanical refiner. In this process the raw material is heated while it is shredded by disk refiners. At this point, chemical additives may be used to improve the strength or other physical properties of the hardboard. The wood fibers are then formed into sheets in the wet-felting process. This process is similar to the Fordrinier process used in paper making. The fibers are first mixed with water in the cyclone and

Table 10
ANNUAL QUANTITY OF SOLID WASTES GENERATED BY STATE IN THE LUMBER AND WOOD PRODUCTS MANUFACTURING INDUSTRY[2-12]

State	1973	1974	1975	1976	1977	1978	1979	1980	1981	1982	1983
Alabama	19.8	604.2	122.4	81.3	162.5	—	—	—	—	56.4	89.9
Arizona	—	0.8	0.6	3.6	66.2	—	—	—	—	—	0.0
Arkansas	130.5	87.2	272.5	400.9	14.3	—	—	—	—	96.8	0.0
California	282.0	326.0	654.0	731.0	1088.0	1357.0	—	394.0	470.0	85.0	481.1
Colorado	68.0	207.0	0.6	257.3	15.1	—	—	—	—	—	0.0
Florida	18.4	29.6	37.7	—	—	—	—	—	—	—	21.6
Georgia	12.1	48.3	384.6	267.8	22.5	—	—	—	—	36.3	54.2
Hawaii	1.6	0.3	—	—	1.3	—	—	—	—	—	0.0
Idaho	228.9	893.0	546.7	412.2	78.2	—	—	—	—	222.4	0.0
Illinois	20.0	21.5	22.4	41.2	64.9	—	—	—	—	12.9	18.8
Indiana	165.9	66.2	32.9	48.0	22.1	—	—	—	—	50.4	86.2
Kansas	2.6	4.6	2.7	2.7	2.7	—	—	—	—	—	0.0
Kentucky	74.3	22.5	20.4	25.1	22.5	—	—	—	—	26.5	20.8
Louisiana	669.3	740.4	695.7	731.0	204.8	—	—	—	—	—	123.8
Maine	18.9	88.3	72.4	126.5	262.5	—	—	—	—	—	0.0
Maryland	0.4	0.5	1.0	0.7	0.7	—	—	—	—	—	0.0
Michigan	20.0	16.5	—	25.9	39.2	—	—	—	—	20.6	62.2
Minnesota	34.4	70.0	47.2	48.9	93.0	—	—	—	—	34.7	54.1
Mississippi	69.8	70.0	111.1	636.8	445.3	411.4	—	386.8	1094.7	188.3	193.8
Missouri	113.9	158.8	223.4	127.9	—	—	—	—	—	—	0.0
Nebraska	0.2	0.2	0.5	2.4	2.6	—	—	—	—	—	0.0
New Hampshire	1.2	0.4	16.9	25.3	3.0	—	—	—	—	—	0.0
New Mexico	7.9	67.2	—	34.2	9.0	—	—	—	—	—	0.0
New York	12.0	43.2	1.5	9.4	12.7	—	—	—	—	—	0.0
North Carolina	89.7	101.5	157.9	92.4	67.4	—	—	—	—	49.4	22.2
Ohio	—	11.0	15.9	15.1	17.5	—	—	—	—	255.4	83.1
Oregon	619.0	921.9	1323.8	1304.0	1057.0	1208.2	889.2	687.2	728.6	797.1	1181.7
Pennsylvania	17.7	8.8	165.5	171.7	195.3	—	—	—	—	47.8	61.6
South Carolina	8.6	5.3	5.2	5.4	124.1	—	—	—	—	42.6	59.0
South Dakota	—	0.8	—	1.4	0.1	—	—	—	—	—	0.0
Tennessee	93.2	13.8	153.3	10.9	51.3	—	—	—	—	—	0.0
Texas	84.8	216.0	172.6	151.0	114.1	—	176.3	—	—	120.0	141.6
Utah	1.3	—	—	7.7	1.6	—	—	—	—	—	0.0
Vermont	27.6	27.5	20.9	25.5	1.4	—	—	—	—	—	0.0

Virginia	535.0	400.8	386.1	383.3	50.6	—	—	—	—	—	73.9
Washington	821.7	686.0	1978.8	2749.4	1528.4	1394.7	1421.0	—	1093.6	720.0	516.3
West Virginia	—	112.2	79.4	79.4	0.9	—	—	—	—	—	0.0
Wisconsin	71.2	84.2	39.9	27.4	64.1	—	—	—	—	79.3	79.8

Note: Values given in thousands of short tons.

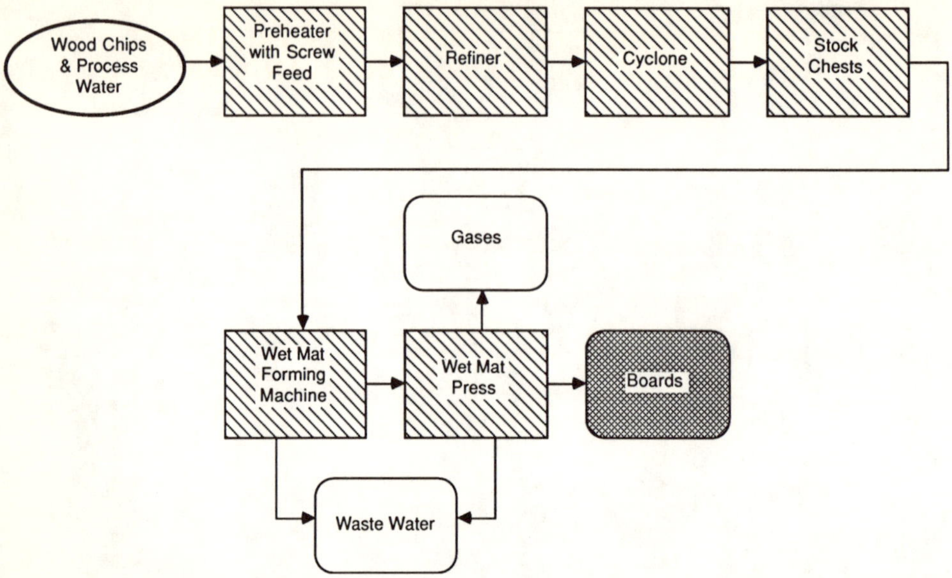

FIGURE 5. Schematic of hardboard manufacturing process.

stock chests until the resulting stock contains about 0.5% fibers by weight. The stock then flows evenly over a screen in the wet forming machine. Water initially is removed by gravity, then by suction, and finally by pressure. The final product, wetlap, is 50 to 75% water. The wetlap is trimmed and cut and delivered to the hardboard press. In the hardboard press a combination of heat and pressure is used to weld the fibers together. The water is removed by the pressure and by evaporation.

Finally, the board may go through a number of finishing processes such as oil tempering, applying paint, or adding wood grain finishes. Oil tempering involves dipping or roller coating the hardboard with drying oils.[20]

C. Disposal and Reuse Practices

1. Wood Preservation Processing Wastes

Solid wastes from the wood preservation industry include tank bottoms and wastewater treatment sludges. Table 11 summarizes the metal concentrations in wood preserving plant sludges. The metal concentrations are higher in the aerated lagoon sediments than the final pond sediments. For example, zinc was between 280 and 310 mg/kg in the aerated lagoon sediments and 48 mg/kg in the final pond sediments.[20] The composition and volume of these sludges depend largely on the processes being used. Sometimes the wastewaters are not treated but used for irrigation.[20] Typically, wastewater sludges are landfilled.

2. Hardboard Manufacturing Wastes

The primary solid wastes from wet-process hardboard production are wastewater treatment sludges which contain wood fibers, soluble wood extractives, zinc and copper from pipe corrosion, and nickel. Metal content of these sludges is included in Table 12. Some of the chemical additives used in manufacturing of hardboards may also be found in these sludges. Additives include vegetable oils, ferric sulfate, aluminum sulfate, petrolatum, thermoplastics, and paints.[21] These sludges are typically landfilled. There are no known reuses for these sludges.

Table 11
METALS ANALYSIS OF WASTEWATER TREATMENT SLUDGES FROM WOOD PRESERVING PLANTS (SIC 2491)[20]

Sludges — Dry Weight (mg/kg)[a]

	Plant #1				
Metals	Aerated lagoon sediments	Final pond sediments	Plant #2 aerated lagoon sediments	Plant #4 estimated by plant mass balance[20]	Plant #5 estimated by plant mass balance[a]
Antimony	<3.70	<1.1	0.990	8.6	—
Arsenic	9.3	7.6	51.0	8.6	6.2
Beryllium	<1.9	3.2	2.5	—	—
Cadmium	2.1	<1.3	0.200	—	—
Chromium	5.6	3.1	56.0	6,200	—
Copper	40.0	4.3	99.0	26.0	28.0
Lead	<2.1	27.0	20.0	8.6	6.2
Mercury	0.136	0.003	0.020	—	—
Nickel	19.0	18.0	18.0	8.6	9.3
Selenium	<4.5	<1.6	0.940	26.0	3.1
Silver	<2.1	0.690	0.910	—	—
Thallium	6.3	0.690	0.910	—	3.1
Zinc	310.0	48.0	280.0	—	605.0

[a] Only used organic preservatives.

Table 12
ESTIMATED METAL CONCENTRATIONS IN HARDBOARD PRODUCTION SLUDGE[21]

Metals	Hardboard sludge (mg/kg[a])
Beryllium	0.11
Cadmium	0.11
Copper	650
Lead	7.2
Nickel	270.0
Silver	0.11
Zinc	980.0

[a] Concentrations estimated from difference in metals from raw wastewater and effluent and from the generated sludge volume.

VI. PAPER AND ALLIED PRODUCTS (SIC Code 26)

A. Waste Generation

The demand for paper and allied products has increased steadily over the last couple of decades as an increasing number of paper products are used in households, in packaging goods, and in work environments. Since 1973, solid wastes generated have averaged about 10 million tons annually with peak generation of 13.25 million tons in 1979 and 13.66 in 1983. The paper and allied products industries generated about 15% of the total manufacturing solid wastes in 1983. Table 13 gives the data on solid wastes generated by state and by year. The largest quantitites of solid waste in the paper manufacturing industry are generated in Illinois, Georgia, Pennsylvania, and Wisconsin. The principal components of the industry are pulp mills, paper mills, and paperboard container and box producing establishments.

B. Manufacturing Processes

Operations in the pulp mills (Figure 6) begin with the preparation of the raw material for processing. Preparation includes washing, debarking, and chipping the logs. Solid wastes resulting from preparation include bark refuse, wood particles, and saw dust.

The wood chips or logs are then chemically or mechanically pulped. In this process, the wood is converted to fibers suitable for paper making; e.g., in stone ground wood, a type of mechanical pulping, debarked logs or billets are fed to grinders where the pulp is ripped from the outsides of the logs. Mechanical pulping typically yields 90% of the raw material as pulp.[21]

Mechanical pulping produces a low-grade paper whereas chemical pulping produces a high-grade paper. Examples of chemical pulping include alkaline pulping and sulfite pulping. Of chemical pulp, 80% is treated using the Kraft process — an alkaline process. In this process, the wood chips are cooked in a caustic soda and sodium sulfite solution. The cooked wood chips are put into a tank where they are separated into fibers. The pulp is washed to remove the residual cooking solution. Finally the pulp is diluted, screened, and deckered. In addition to wastewater, these processes produce knots and fiber residues.

Pulping is followed by bleaching. Mechanical pulp is typically bleached with hydrosulfides and peroxides. Chemical pulp is most commonly bleached with chlorine, calcium, or sodium hypochlorite and chlorine dioxide. The bleached pulp is now ready to be manufactured into paper. First the pulp is refined. In this process, water is added to the bleached pulp to form a slurry which is treated in beaters or refiners. This process improves the mat-forming properties of the fibers. It is a mechanical process and does not produce much solid waste.

The pulp slurry is mixed with paper-making additives. Additives include mineral pigments used as filler, resin used to increase the liquid resistance of the paper, and dyes to color the paper. The pulp is then formed into sheets using the fourdrinier paper machine or a cylinder machine. The pulp slurry is fed through screens into the machine where sheets are formed. The water is removed by pressing and drying.[21]

After dewatering of the sheets, coatings containing mineral pigments and adhesives are added. The coatings improve the surface of the paper for printing. The coatings are applied mechanically by coating machines.[22]

C. Disposal and Reuse Practices

Paper manufacturing uses large quantities of water which results in large volumes of wastewaters. Treatment of the wastewater produces primary and secondary sludges (Table 14). It is estimated that 300 kg of solid waste are produced for every ton of finished paper produced.[22] The primary sludges contain fibers, filler materials, and other chemical additives. Typically, these sludges are disposed of in industrial landfills.

VII. CHEMICALS AND ALLIED PRODUCTS (SIC Code 28)

The chemicals and allied products group is one of the largest and most important manufacturing industry groups in the U.S. In 1983 this industry group generated over

Table 13
ANNUAL QUANTITY OF SOLID WASTES GENERATED BY STATE IN THE PAPER AND ALLIED PRODUCTS MANUFACTURING INDUSTRY[2-12]

State	1973	1974	1975	1976	1977	1978	1979	1980	1981	1982	1983
Alabama	222.0	336.2	582.1	415.3	647.6	651.7	568.6	744.2	802.0	791.9	650.2
Arkansas	132.3	—	363.8	327.9	600.8	601.6	343.8	324.0	299.1	372.2	0.0
California	138.0	234.0	126.0	186.0	218.0	243.0	240.0	254.0	214.0	162.0	436.2
Connecticut	94.5	91.2	65.6	70.0	41.3	—	—	—	—	33.5	16.8
Florida	350.7	834.4	382.6	489.6	168.0	115.0	147.1	199.7	164.4	122.4	147.4
Georgia	216.7	747.9	761.8	835.1	1047.1	961.1	721.5	1029.8	1017.2	1037.2	1050.5
Illinois	691.3	215.8	195.3	202.9	161.7	194.4	203.6	250.4	170.7	55.8	182.7
Indiana	17.1	23.1	19.0	27.5	58.4	—	—	—	—	46.9	46.5
Kansas	44.4	30.5	71.4	72.5	9.3	—	—	—	—	—	0.0
Kentucky	128.2	146.9	143.9	116.5	95.4	93.4	—	—	—	—	40.8
Louisiana	115.2	88.6	171.2	368.5	476.5	554.3	252.9	194.5	220.0	211.9	308.2
Maine	12.7	526.1	299.6	287.6	491.8	531.1	490.8	521.0	735.0	904.8	1261.6
Maryland	—	—	—	—	132.0	136.9	—	126.3	125.1	—	7.7
Massachusets	65.1	59.2	116.4	172.6	157.4	180.0	—	124.7	135.2	53.1	116.3
Michigan	691.3	398.5	329.5	347.7	367.6	—	727.1	—	—	398.2	388.7
Minnesota	253.3	231.8	220.8	247.9	250.2	235.4	126.8	156.3	157.2	—	222.4
Missouri	29.9	18.8	26.8	30.6	16.8	—	—	—	—	—	19.8
Nebraska	—	3.1	11.0	2.4	—	—	—	—	—	—	0.0
New Jersey	112.6	181.3	75.0	92.7	76.5	72.1	—	—	367.1	464.1	110.0
New York	215.0	403.3	325.8	361.0	275.4	294.5	263.5	205.5	375.0	466.8	311.0
North Carolina	523.7	596.9	700.5	700.8	820.5	945.0	908.0	886.1	310.6	794.6	0.2
Ohio	746.1	678.5	631.7	791.9	502.2	507.7	530.1	670.9	610.7	695.6	438.3
Oregon	126.4	137.9	128.7	292.2	418.8	361.2	328.6	194.0	300.4	145.7	291.1
Pennsylvania	501.8	369.6	406.1	543.3	462.3	495.2	521.9	517.0	610.9	624.6	1178.5
Rhode Island	9.7	9.1	4.3	3.0	1.2	—	—	—	—	—	0.0
South Carolina	32.3	126.6	183.6	229.8	264.9	—	410.2	—	—	299.8	329.8
South Dakota	—	0.1	—	0.4	—	—	—	—	—	—	0.0
Tennessee	82.6	132.8	141.1	146.8	171.8	167.6	214.6	167.8	176.5	—	199.6
Texas	94.5	122.7	220.5	308.3	254.0	272.1	347.3	402.8	386.7	675.0	659.0
Utah	6.0	0.3	—	4.4	5.8	—	—	—	—	—	0.0
Virginia	285.6	281.7	352.5	341.1	457.3	423.2	480.9	454.3	552.9	434.0	745.9

Table 13 (continued)
ANNUAL QUANTITY OF SOLID WASTES GENERATED BY STATE IN THE PAPER AND ALLIED PRODUCTS MANUFACTURING INDUSTRY[2-12]

State	1973	1974	1975	1976	1977	1978	1979	1980	1981	1982	1983
Washington	229.0	146.7	229.2	383.2	364.2	306.6	546.7	690.4	638.7	511.4	614.0
West Virginia	8.9	2.8	—	3.2	5.8	—	—	—	—	—	0.0
Wisconsin	483.7	865.1	1189.0	1001.5	892.0	1035.3	—	1104.9	1198.7	1137.0	1300.0

Note: Values given in thousands of short tons.

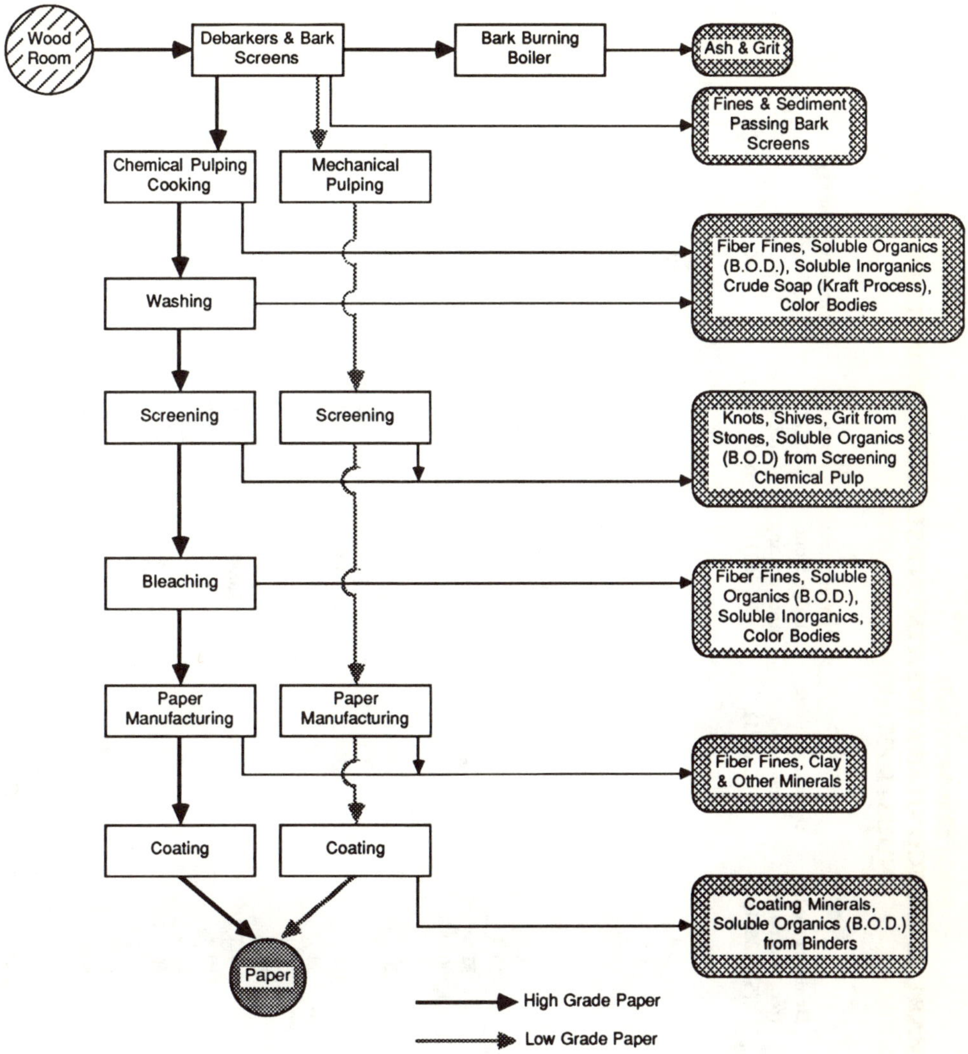

FIGURE 6. Schematic diagram of paper production.

21% of all manufacturing industry solid waste — the largest quantity of solid waste by a single manufacturing group. Table 15 gives quantities of solid wastes generated by state and by year.[2-12] The chemical industry solid waste generation peaked in 1975 and has been declining ever since. Despite this trend, it still generates about 20% of all manufacturing industry solid waste. The largest volumes of waste are produced in Louisiana and Texas.

Not only is the chemical industry a major generator of solid waste, the majority of this solid waste is classified as hazardous. According to JRB Associates,[23] in 1980 the chemicals and allied products industry generated 28.1 million short tons of hazardous waste. That year the industry generated 43.4 million short tons of solid waste, so, about 65% of the solid waste generated was classified as hazardous. In addition, of the manufacturing industries included between SIC 22 and SIC 39, the chemical and allied products industry generated about 62% of the total hazardous waste generated by these manufacturing industries.

Table 14
CHEMICAL ANALYSIS OF PRIMARY AND SECONDARY TREATMENT SLUDGES FROM THE PAPER INDUSTRY[21]

Constituent	Primary and secondary sludge from semi-chemical pulping (SIC 2611)	Deinking sludge (recycled paper) (SIC 2621)	Deinking sludge new paper (SIC 2621)	Pretreatment sludge from paper coating (SIC 2631)	Board mill sludge (SIC 2631)	Combined primary and secondary sludge (SIC 26)
Water (%)	90—96	77.06	—	86	—	40
Solids (%)	4—10	22.4	—	14	—	60
Ash (%)	1—2.5	—	—	—	—	40
COD	60,000—120,000	—	—	400,000	—	—
Phenol	5	—	—	—	—	—
PCB	<13	—	—	—	—	—
Oil and grease	1	—	—	—	—	—
Total nitrogen	1,400	—	—	—	—	4
Aluminum	—	21,300	—	100,000	—	—
Cadmium	1.5	32	—	—	—	4.6
Calcium	4,000—15,000	4,390	—	—	—	—
Chloride	—	332	—	—	—	—
Chromium	—	20	180	—	79	16
Copper	N.D.[a]	—	330	—	62	—
Iron	—	538	1,500	200,000	2,400	—
Lead	120	32	1,300	—	380	47
Magnesium	—	1,170	—	—	—	—
Manganese	250	16	—	—	—	1,146
Nickel	25	2.3	8	3,000	11	52
Phosphorous	—	310	—	—	—	2
Potassium	1,600	114	—	—	—	—
Sodium	1,400	146	—	—	—	—
Sulfate	120	0.03	—	—	—	—
Zinc	260	151	300	4,000	350	397

Note: Values given in ppm unless otherwise noted.

[a] = N.D., not detected.

Table 15
ANNUAL QUANTITY OF SOLID WASTES GENERATED BY STATE IN THE CHEMICALS AND ALLIED PRODUCTS MANUFACTURING INDUSTRY[2-12]

State	1973	1974	1975	1976	1977	1978	1979	1980	1981	1982	1983
Alabama	74.5	465.2	542.2	556.5	621.6	455.2	463.9	460.7	459.3	371.4	115.8
Alaska	—	1.2	1.2	—	—	—	—	—	—	—	115.8
Arkansas	24.6	1,408.1	1,074.0	1,075.1	1,028.3	—	—	—	—	—	0.0
California	477.0	329.0	514.0	321.0	447.0	471.0	369.0	544.0	340.0	349.0	338.8
Connecticut	85.8	92.0	77.3	79.7	104.0	101.4	101.0	89.6	84.1	74.5	83.1
Delaware	—	154.6	127.4	208.6	161.8	181.7	212.3	219.4	235.8	207.7	241.1
Florida	1,227.4	18,882.7	13,124.4	22,011.3	26,390.3	19,415.2	13,445.2	12,217.7	12,080.8	12,723.0	133.1
Georgia	113.1	169.8	122.4	77.5	47.6	54.2	69.9	52.7	50.9	61.8	189.6
Idaho	—	—	—	—	3,192.8	2,866.8	—	—	—	—	0.0
Illinois	37.0	299.1	374.4	412.2	1,338.9	1,779.0	1,785.8	1,893.3	1,859.4	1,114.6	411.5
Indiana	189.5	223.0	276.5	400.2	357.2	456.5	324.6	352.5	196.5	296.6	493.5
Iowa	17.3	675.0	4,300.8	48.2	652.6	648.8	—	63.8	58.3	29.9	115.6
Kansas	16.5	15.0	14.9	26.2	41.2	36.3	—	39.9	43.2	47.2	61.8
Kentucky	111.2	177.7	227.7	231.2	294.9	324.4	297.1	201.0	268.6	366.5	230.4
Louisiana	341.2	1,588.3	1,813.6	6,155.4	8,098.2	7,897.9	8,232.3	8,546.7	7,539.1	5,604.7	2,946.1
Maine	—	100.6	80.5	2,323.2	47.1	—	—	—	—	59.6	40.2
Maryland	38.6	61.6	59.6	55.4	187.4	211.2	178.4	194.1	392.7	—	278.9
Massachusetts	74.5	85.5	76.2	77.0	72.1	73.3	78.0	71.7	119.9	322.1	479.7
Michigan	37.0	432.9	299.6	610.2	666.4	611.8	595.2	787.8	783.8	452.9	443.4
Minnesota	—	—	—	—	30.6	—	35.2	—	—	2.0	9.1
Mississippi	1,122.9	1,151.6	1,123.4	1,157.5	1,167.9	1,165.6	—	—	—	678.2	3.9
Missouri	79.8	96.7	124.6	102.8	98.1	130.4	150.7	136.2	112.9	63.6	71.2
Nevada	—	2.9	18.7	24.5	16.0	—	—	—	—	—	0.0
New Hampshire	2.6	1.3	1.3	0.4	—	—	—	—	—	—	0.0
New Jersey	479.6	572.7	547.8	479.4	428.6	468.4	684.1	691.1	692.3	643.2	658.4
New York	126.7	727.1	716.1	719.7	813.8	865.0	880.3	881.2	910.8	330.4	426.4
North Carolina	203.1	296.4	201.6	300.4	415.1	407.9	630.3	604.9	651.9	689.4	412.1
Ohio	310.4	369.3	1005.0	1,380.1	343.0	474.6	435.7	445.0	457.1	1,046.5	509.9
Oklahoma	0.7	57.8	57.8	101.4	6.1	—	—	—	—	33.7	52.2
Pennsylvania	183.9	317.3	291.8	304.5	304.3	329.7	313.9	267.4	309.7	349.1	432.2
South Carolina	63.8	111.0	145.7	149.0	241.1	243.9	353.9	235.5	281.5	—	197.5

Table 15 (continued)
ANNUAL QUANTITY OF SOLID WASTES GENERATED BY STATE IN THE CHEMICALS AND ALLIED PRODUCTS MANUFACTURING INDUSTRY[2-12]

State	1973	1974	1975	1976	1977	1978	1979	1980	1981	1982	1983
Tennessee	369.0	534.9	1,502.6	1,176.5	1,542.9	1,671.3	2,121.3	2,717.6	1,676.0	1,459.3	639.2
Texas	3,701.1	5,400.4	6,940.0	7,288.3	5,317.7	5,208.4	6,497.6	5,871.8	6,830.3	3,864.9	3,745.2
Utah	3.1	933.9	849.0	0.4	52.0	—	—	—	—	—	0.0
Virginia	152.8	303.6	335.1	307.2	258.7	283.9	382.7	402.3	457.9	469.2	599.9
Washington	—	14.4	14.2	12.8	139.0	—	—	—	—	7.0	6.3
West Virginia	—	467.2	336.0	376.4	452.2	432.6	432.4	310.8	330.8	509.8	979.9
Wisconsin	—	13.3	26.7	30.2	48.7	—	30.5	—	—	28.7	26.5
Wyoming	65.7	0.5	0.4	—	20.7	—	—	—	—	—	0.0

Note: Values given in thousands of short tons.

The chemical and allied products group includes many different industries. Major manufacturing subgroups include:

- SIC 281 Industrial inorganic chemicals
- SIC 282 Plastic materials and synthetics
- SIC 283 Drugs
- SIC 284 Soaps, cleaners, and toilet goods
- SIC 285 Paints and allied products
- SIC 286 Industrial organic chemicals
- SIC 287 Agricultural chemicals
- SIC 289 Miscellaneous and chemical products

All these industries manufacture chemicals, but the raw materials, the unit processes, and consequently the characteristics and quantities of the solid wastes vary between and within the subgroups. A more detailed discussion of the manufacturing processes and solid waste disposal and reuse practices is included in Chapter 7.

A. Inorganic Chemicals

The inorganic chemicals industry (SIC 281) includes plants that manufacture alkalies and chlorine, industrial gases, inorganic pigments, and other inorganic compounds. The alkalies and chlorine producing facilities primarily manufacture chlorine, potassium carbonate, sodium hydroxide, potassium hydroxide, and potassium bicarbonate. Industrial gas plants manufacture acetylene, carbon dioxide, argon, helium, hydrogen, nitrogen, oxygen, nitrogen oxide, and rare gases. Production of inorganic pigments yields products such as titanium pigments, zinc oxide pigments, zinc sulfide, chrome colors, chrome yellow and orange, zinc yellow, iron oxide pigments, and red lead. Other industrial inorganic chemicals subgroups include manufacture of nitric acid, ammonia, ammonium nitrate, ammonium sulfate, calcium hypochlorite, sulfuric acid, boric acid, chromic acid, hydrochloric acid, hydrofluoric acid, aluminum oxide, aluminum chloride, aluminum hydroxide, and aluminum sulfate.

The inorganic chemicals industry in 1983 generated about 32% of the chemical industry solid waste.[12] The solid waste from the inorganic chemicals industry is difficult to characterize because these characteristics are directly related to the chemicals being produced. To illustrate this point, in Table 16 effluent from a plant manufacturing sodium sulfite, sodium thiosulfate, and zinc sulfate is compared to the effluent from a plant manufacturing soda ash, sodium silicate, and calcium carbonate. The properties of these effluents are completely different, yet both of these are industrial inorganic chemical manufacturers.[17]

B. Plastics and Synthetics

The plastics and synthetics subgroup (SIC 282) has seen phenomenal growth in the last two decades. The primary products manufactured are resins, cellulose, plastic, synthetic rubber, rayon and similar cellulose fibers, and synthetic organic fibers such as nylon and Acrilan.®

The plastics and synthetics group in 1983 generated about 11% of the chemical industry solid waste.[12] The wastewaters and wastewater treatment sludges from this group inevitably contain organics. The concentrations and types of organics depend on the manufacturing process used. As an example, chemical data for treatment sludges from two synthetic rubber manufacturing facilities (SIC 2822) are summarized in Table 17. These wastes contain trace concentrations of organics and metals. Solid waste from this industry is disposed of on land or incinerated.

Table 16
COMPARISON OF EFFLUENTS FROM TWO INDUSTRIAL INORGANIC CHEMICAL MANUFACTURERS[17]

Characteristics	Manufacturer 1	Manufacturer 2
Products manufactured	Sodium sulfite, sodium thiosulfate, and zinc sulfate	Soda ash, sodium bicarbonate, chlorine, caustic soda, hydrochloric acid, sodium silicate, and calcium carbonate
Effluent flow rate (gallons per day)	300,000 for process, 1,500,000 total	6,000,000 total
Effluent pH	6.5—8.5	11
SO_3 (ppm)	20,000	—
Zn	20	—
NaCl (ppm)	15,000	—
$NaHCO_3$ (ppm)	15,000	—
Suspended solids (ppm)	—	30—50
Cl (ppm)	—	79,000
NH_3 (ppm)	—	4
Total soluble solids (ppm)	—	131,000
Fe soluble (ppm)	—	0.36
Phenols (ppb)	—	219
Hardness (equiv. $CaCO_3$ mg/ℓ)	—	76,000

Table 17
CHEMICAL COMPOSITION OF SYNTHETIC RUBBER MANUFACTURING TREATMENT SLUDGES[19]

Parameter	Plant #1 (mg/ℓ)	Plant #2 (mg/ℓ)
Metals		
Arsenic	0.017	0.220
Selenium	<0.010	<0.010
Cadmium	0.002	0.067
Beryllium	<0.003	<0.016
Copper	0.410	2.10
Antimony	<0.010	<0.010
Chromium	1.300	28.0
Nickel	0.036	3.50
Zinc	0.250	150
Silver	<0.010	0.014
Thallium	0.010	0.010
Lead	0.070	2.60
Mercury	0.004	0.005
Organics		
Total phenols	0.020	0.139
2-Nitrophenol	—	0.052
Phenol	1.80	0.015
2-Chlorophenol	0.207	0.018
2,4,6-Trichlorophenol	0.072	0.043
2,4-Dinitrophenol	N.D.	0.483
Pentachlorophenol	N.D.	—
4-Nitrophenol	0.133	0.485
Phenanthrene and anthracene	0.110	—
Naphthalene	0.010	—
Pyrene	0.012	—

Note: N.D., Not detected.

Table 18
CHEMICAL ANALYSIS OF WASTEWATER TREATMENT SLUDGES GENERATED BY THE PHARMACEUTICALS INDUSTRY[19]

Parameter	Plant #1 sludge (mg/ℓ)	Plant #2 clarifier sludge (mg/ℓ)
Metals		
Arsenic	<0.010	0.021
Selenium	0.033	<0.101
Cadmium	0.059	0.016
Beryllium	<0.003	<0.003
Copper	8.300	2.800
Antimony	<0.010	<0.010
Chromium	0.740	2.10
Nickel	0.110	0.032
Zinc	7.700	7.800
Silver	0.160	0.072
Thallium	<0.010	0.056
Lead	0.810	0.030
Mercury	0.062	<0.010
Organics		
Total cyanides	<0.050	0.230
Total phenols	0.074	0.201
Phenol	0.009	—
Pentachlorophenol	0.038	—

C. Pharmaceuticals Industry

The pharmaceutical industry (SIC 283) manufactures medicines. There are six basic processes used by this industry: fermentation, synthesis, formulation, fractionation, extraction, and the growth of biological cultures.[19] In 1983 this industry generated about 6% of the chemical industry solid waste.[12] Chemical data for treatment sludges from two pharmaceutical plants are included in Table 18. Metal and organic concentrations in these sludges are very low. In addition to wastewater treatment sludges, other solid wastes include spent chemicals, product tailings, product rejects, biological wastes, and packaging materials. Solid wastes are disposed of in landfills, while combustible wastes are often incinerated.

D. Soaps, Cleaners, and Toilet Goods

This subgroup (SIC 284) is a very small generator of solid waste. In 1983 it generated about 2% of the chemical industry solid waste.[12] This is only about 0.4% of the total solid waste generated by all manufacturing industries combined.

E. Paints and Allied Products

The paints and allied products manufacturing industry (SIC 285) is distributed fairly uniformly across the U.S. and manufactures paints, varnishes, lacquers, enamels, shellac, putty, caulking compounds, food fillers and sealers, paint and varnish removers, etc. Production in this industry closely follows the housing construction industry. The paints and allied products subgroup generates less than 1% of the chemical industry solid wastes. The paint formulating sludges, however, contain metals and organic compounds (Table 19).

Table 19
CHEMICAL ANALYSIS OF PAINT FORMULATING INDUSTRY SLUDGES[19]

Parameter	Median (ppm)	Minimum (ppm)	Maximum (ppm)
Inorganic constituents			
Aluminum	600,000	50,000	3,000,000
Barium	4,345	72	50,000
Boron	2,000	175	14,200
Calcium	914	250	30,200
Cobalt	600	<50	15,600
Iron	200,000	22,600	8,000,000
Magnesium	79	<1	1,500
Manganese	5,000	300	50,000
Molybdenum	1,000	<50	15,000
Sodium	260	100	35,000
Tin	2,000	200	14,500
Titanium	20,000	1,000	230,000
Vanadium	400	60	11,500
Yttrium	<200	45	<600
Antimony	150	<10	13,000
Arsenic	495	<25	2,000
Beryllium	20	2	3,760
Cadmium	200	<8	14,700
Chromium	700	<50	90,000
Copper	1,000	216	80,000
Cyanide	<20	<1	36,500
Lead	3,000	100	80,000
Mercury	640	5	220,000
Nickel	<200	<20	<200,000
Selenium	<250	8	<2,000
Silver	<10	<2	<100
Thallium	<400	<10	<2,000
Zinc	90,000	<600	2,000,000
Organics			
Benzene	30	<10	1,900
Carbon tetrachloride	<10	<10	<10
Chlorobenzene	176	12	340
1,2-Dichlorethane	17	17	17
1,1,1-Trichloroethane	14	<10	3,200
1,1,2,2-Tetrachloroethane	13	<10	17
2,4,6-Trichlorophenol	10	<10	10
Chloroform	920	840	1,000
2,4-Dimethylphenol	<10	<10	<10
Ethylbenzene	237	26	99,000
Methylene chloride	1,735	300	900,000
Naphthalene	202	<10	1,050
2,4-Dinitrophenol	<18	10	27
Pentachlorophenol	125	35	1,100
Phenol	150	<10	1,120
Di(2-ethylhexyl) phthalate	215	<10	1,940
Butyl benzyl phthalate	1,412	18	38,800
Di-n-butyl phthalate	70	<10	17,750
Diethyl phthalate	100	50	760
Dimethyl phthalate	<10	<10	<10
Anthracene	<210	<10	410
Pyrene	<10	<10	<10
Tetrachloroethylene	710	<10	8,200
Toluene	905	130	350,000

Table 19 (continued)
CHEMICAL ANALYSIS OF PAINT FORMULATING INDUSTRY SLUDGES[19]

Parameter	Median (ppm)	Minimum (ppm)	Maximum (ppm)
Trichloroethylene	<10	<10	130
Aldrin	<10	<10	<10
Beta-endosulfan	<10	<10	<10
Delta-BHC	<10	<10	<10

Note: Paint formulating industry sludges produced from physical-chemical treatment and/or settling.

F. Organic Chemicals

The industrial organic chemicals subgroup (SIC 286) consists of facilities that produce organic chemicals from hydrocarbons. This subgroup started its activities using hydrocarbons obtained from coal. Since World War II, the hydrocarbons from petroleum refining have been used. Because the raw material hydrocarbons are produced by the same plant which uses them for final products, it is not easy to separate the waste production in the petrochemical or the coal tar products industry from that of organic chemicals. During the last 10 years, the industrial organic chemicals manufacturing group has produced about 4 million tons of solid wastes annually.[2-12] In 1976, a peak of 8.4 million tons occurred. In 1983, organic chemical production generated about 4% of the total of manufacturers' solid wastes.[12]

G. Agricultural Chemicals

Agricultural chemical plants (SIC 297) manufacture fertilizers, herbicides, pesticides, and other agricultural chemicals. Fertilizers are mostly mixtures of nitrogen, phosphorous, and potash compounds. Pesticides include halogenated organics, organophosphates, and organonitrates. The agricultural chemicals industry generates about 15% of the chemical industry solid waste. The solid wastes are generated either directly as residuals from mixing operations and spills, or as sludges from settling ponds. Typically, solid wastes are rich in the materials being manufactured. For example, wastewater treatment sludges from a toxaphene plant were 19% toxaphene.[19] Disposal methods include incineration, landfilling, and discharge of treated liquids to municipal sewage systems. Wastes are also disposed of in ponds or applied to soils as part of land treatment systems.

VIII. PETROLEUM AND COAL PRODUCTS (SIC Code 29)

A. Waste Generation

The petroleum and coal products industries are a large component of the U.S. manufacturing industries, yet these industries do not generate a proportionately large amount of solid waste. Table 20 lists the quantities of solid wastes generated by state and year for this industry group.[2-12] Solid waste generation averaged about 3.2 million tons per year during the 1973 to 1983 time period with peak production of 4.9 million tons in 1980. In 1983, the petroleum and coal products industry generated about 4% of the total manufacturing solid wastes. Table 20 indicates that Texas is the largest generator of petroleum and coal products' solid wastes. As an example, the manufacturing processes and disposal/reuse practices in the petroleum refining industry and rerefining industry are discussed below.

Table 20

ANNUAL QUANTITY OF SOLID WASTES GENERATED BY STATE IN THE PETROLEUM AND COAL PRODUCTS MANUFACTURING INDUSTRY[2-12]

STATE	1973	1974	1975	1976	1977	1978	1979	1980	1981	1982	1983
Arizona	—	0.8	0.2	0.4	—	—	—	—	—	—	0.0
California	199.0	322.0	285.0	278.0	284.0	640.0	429.0	561.0	429.0	507.0	449.4
Delaware	—	—	—	—	—	—	—	14.0	—	9.2	0.0
Florida	15.3	6.6	8.6	8.9	19.1	—	—	—	—	11.7	18.1
Georgia	36.9	26.5	33.9	38.8	45.0	—	—	—	—	—	29.4
Illinois	297.7	305.9	264.9	340.7	228.9	252.9	289.9	238.0	176.6	166.4	146.6
Indiana	50.5	—	—	—	—	184.7	—	—	198.7	84.2	79.5
Kansas	—	—	—	—	—	—	—	118.2	60.3	—	20.3
Louisiana	38.6	81.7	94.5	129.9	212.1	224.1	212.4	791.9	778.6	695.9	323.7
Maryland	—	—	22.9	36.0	27.5	—	—	—	—	—	0.0
Michigan	297.7	—	—	—	—	8.5	—	—	3.6	1.3	0.0
Minnesota	—	—	—	—	22.0	38.5	—	—	42.2	31.2	31.2
Mississippi	—	—	—	—	—	—	—	—	21.6	19.4	0.0
Missouri	—	—	—	—	—	47.0	—	58.5	23.7	43.4	43.7
Montana	—	—	—	—	10.1	10.7	—	—	13.5	6.9	0.0
Nebraska	0.1	0.2	—	0.1	0.1	—	—	—	—	—	0.0
New Jersey	90.7	78.1	223.5	117.8	123.7	77.3	61.7	70.5	125.0	152.8	170.2
New York	—	—	5.8	—	—	—	—	—	—	2.2	10.0
Ohio	58.5	111.0	105.7	214.4	117.4	175.8	119.6	99.7	82.4	73.4	67.1
Oklahoma	41.5	53.5	27.6	174.5	26.1	35.3	—	44.9	69.8	20.0	47.6
Oregon	0.3	0.3	0.3	0.2	80.4	—	—	—	—	—	0.4
Pennsylvania	127.6	109.4	178.0	133.5	110.6	132.3	172.3	246.1	201.6	—	175.1
Rhode Island	—	—	0.1	0.1	—	—	—	—	—	—	0.0
South Carolina	58.4	30.1	61.0	43.2	—	—	—	—	—	—	0.0
Tennessee	17.9	14.1	14.1	8.7	50.9	—	—	—	—	—	0.0
Texas	354.0	706.3	381.4	658.2	925.3	881.7	706.2	2032.8	2002.6	2078.7	1543.9
Washington	—	—	—	—	—	16.7	—	14.4	16.5	13.4	14.2
West Virginia	312.9	—	—	—	—	—	—	—	—	7.3	5.9

Note: Values given in thousands of short tons.

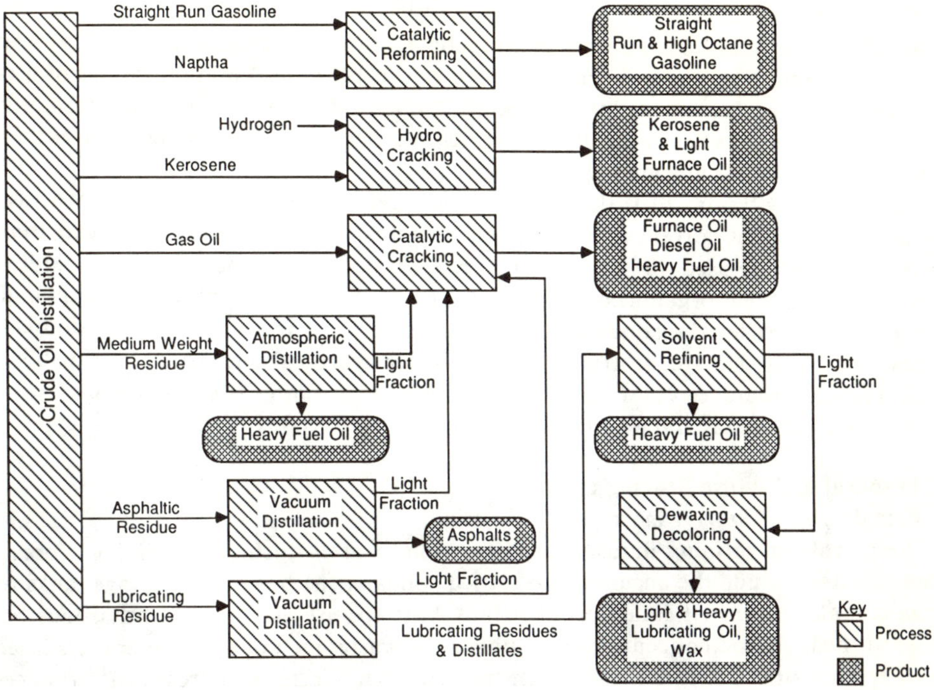

FIGURE 7. Schematic diagram for petroleum refinery.[24]

B. Manufacturing Processes

1. Petroleum Refining

Petroleum refineries (SIC 2911) process crude oil into petroleum products such as gasoline, jet fuel, fuel oil, hydraulic oil, and gases. Refining includes a combination of separation and conversion processes. The specific processes used vary from one refinery to another and may even vary from day to day at a specific refinery, because the processes used depend on the desired petroleum products and the chemical characteristics of the incoming crude oil.

Figure 7 is a process diagram for a generalized refinery. The first process invariably is crude oil distillation where the crude oil is separated by weight into fractions. In this example the crude is initially separated into eight fractions: (1) gas, (2) straight run gasoline, (3) naphtha, (4) kerosene, (5) gas oil, (6) medium weight residues, (7) asphaltic residue, and (8) lubricating residue.[24]

Both the straight run gasoline and naphtha then pass through a catalytic reformer. This process converts the naphthas and other hydrocarbons to aromatics which improves the octane rating of the gasoline. This process produces gas, straight run gasoline, and high octane gasoline.

The kerosene and gas oil fractions are processed in a hydrocracking unit. In this process the temperature of the hydrocarbons is raised in a hydrogen atmosphere. This breaks the hydrocarbons into lighter straight or branch chained molecules which can be used in formulating gasoline.

The residues, asphaltic residues, and lubricating residues are distilled further by either atmospheric or vacuum distillation. The heavy distillates form heavy fuel oil, asphalts, and lubricating oil. The lighter distillates are processed in catalytic cracking units. In this process the temperature of the hydrocarbons is raised in the presence of a catalyst. The hydrocarbons break down into lighter hydrocarbons which are suitable for gasoline, diesel oil, and heavy fuel oils.

2. Rerefining (SIC 2992)

Rerefining (SIC 2992) is another manufacturing process included in SIC 29. It is an alternative to the disposal or burning of used oil. In 1983 106.7 million gallons of used oil were rerefined for use as lubricating oil.[25] Used oil, particularly used automotive oil, typically contains heavy metals and lighter hydrocarbons. Rerefining removes these substances and yields lubricating oil.

The acid/clay treatment is the predominant rerefining technique.[26] Other processes include caustic clay, solvent extraction, and vacuum distillation. In the acid/clay treatment (Figure 8) the used oil is mixed with concentrated sulfuric acid. The acid oxidizes and coagulates the inorganic contaminants. The parafinic and naphthenic compounds are left intact. After the solids are removed, the oil is pumped to a steam and clay contacting vessel. The steam strips the oil of lighter hydrocarbons producing a distillate fuel. The clay removes color bodies and suspended carbon. The residual clay is removed by a filterpress and a suitable lube stock is produced.

C. Disposal and Reuse Practices

1. Petroleum Refining Wastes

The petroleum refining industry generates solid wastes during the refining process. These wastes include the accumulated settled solids at the bottom of storage tanks or process tank bottoms. Leaded product tank bottoms are classified as hazardous because of the high lead concentrations. Other solid wastes include spent filter clays (used to filter impurities from the oil) and spent catalysts. Wastewater treatment produces sludges. Air pollution control also generates scrubber sludges and particulates containing catalysts. Table 21 summarizes the sources of refinery wastes and the types of wastes produced from each. Table 22 summarizes metal concentrations from a number of these solid wastes. Metals high in the sludges include chromium, copper, lead, and zinc. The highest concentrations typically occur in API separator sludge, wastewater treatment sludges, and cooling tower basin sludge.[19]

In order to reduce solid waste disposal, the refinery industry has been instituting resource recovery procedures to reclaim oil and catalysts from the wastes; e.g., the lighter oil phases are reclaimed from refinery wastewaters using the API oil/water separator. Spent catalysts are processed for metal recovery. Spent filter clays are regenerated in multiple hearth furnaces.

A number of methods are used to reduce the volume of wastes before disposal. The sludges are dewatered as much as possible by gravity thickening, centrifugation, and filtration. Incineration reduces sludge volumes by consuming the carbon in the sludge. Incineration still leaves behind small quantities of ash.

The ultimate disposal of solid wastes typically has been in landfills. These solid wastes, because of high hydrocarbon content, are well suited to land farming. Both oily and biological treatment sludges have been applied to the land, where soil bacteria degrade the sludges. Digested biological treatment sludges have also been used as soil conditioners for lawns and shrubbery.[27]

2. Rerefining Industry Wastes

Rerefining produces four major wastes: (1) settled sludges, (2) wastewater, (3) spent clays, and (4) distillation bottoms. Metals and solvent concentration for these materials are summarized in Table 23.[25] Other wastes include tank bottoms, acid sludges, solvent sludges, ultrafiltrate solids, and centrifuge sludges.

Typically these wastes are disposed of in landfills. Acid sludges have been used in brick and paving material, in acid recovery, and as fuel. The spent clay is sometimes stored for later recovery or regeneration.

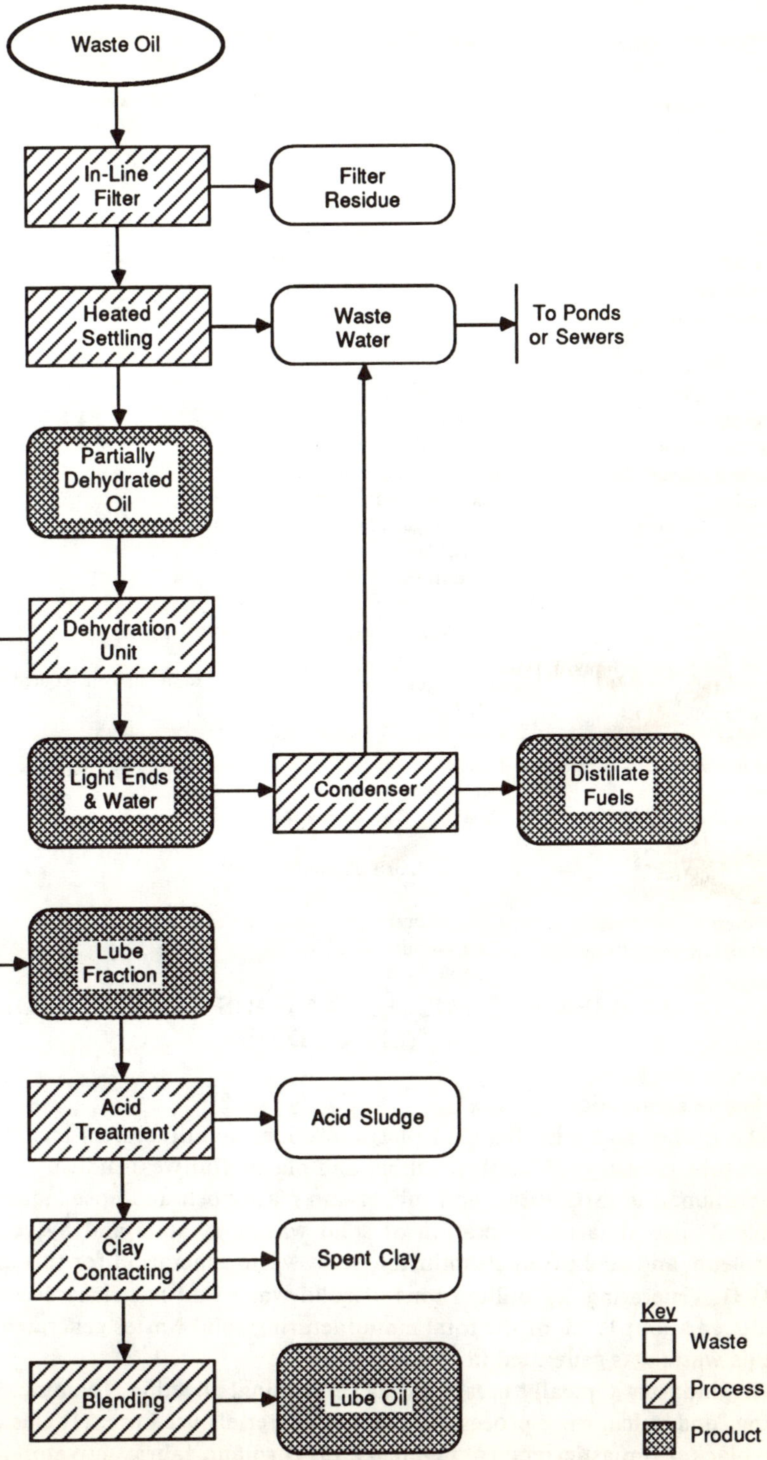

FIGURE 8. Schematic diagram of process to rerefine waste oil into lube oil.[25]

Table 21
SOURCES AND CHARACTERISTICS OF REFINERY SOLID WASTES[26]

Source	Typical solid wastes
	Refining Processes
Tank bottoms	
Crude oil storage	Wax bottoms, sand, rust, and silt
Leaded product storage	Lead sludge, sand, rust, and silt
Crude processing	
Thermal cracking	Separator coke
Catalytic cracking	Spent catalyst
Catalytic reforming	Spent catalyst
Polymerization	Spent catalyst
Alkylation	Corrosion products (sludge, tar)
HF alkylation	Calcium fluoride sludge
Asphalt manufacture	Asphalt drips, asphalt emulsions, emulsions, light solvent
Cooling	Coke fines, wax tailings
Product treating	Acid sludge, adsorbents, lead sludge
Lubes and grease	Soaps, clay
Wax	Slops, drips
	Wastewater, Pollution Control
API separators	Oil sludges, heavy hydrocarbons, sand, silt, oily coated inert solids
Air flotation	Oily froth
Clarifiers	Biological floc, flocs
Flocculation	Floc, clay
Filters	Filter cake
Biological oxidation	Biological floc, biological sludge
	Air Pollution Control
Bag filters	Catalyst dust
Electrostatic precipitators	Catalyst dust

IX. RUBBER AND MISCELLANEOUS PLASTIC PRODUCTS (SIC CODE 30)

A. Waste Generation

The rubber and miscellaneous plastic product manufacturers include the tires and inner tube industry (SIC 3011), rubber and plastic footwear industry (SIC 3021), rubber reclamation (SIC 3031), and rubber and plastic belt and hose industry (SIC 3041). Table 24 lists data on generation of solid wastes by state and by year.[2-12] Like the petroleum and coal products industry, solid waste generation for this industry peaked in 1981, generating 2.9 million tons of solid waste and has since declined. In 1983 it produced about 1.7% of the total manufacturing solid wastes generated. The majority of this waste was generated in Ohio.

This industry typically uses molding, extending, sheeting, foaming, coating, fabricating, and vulcanizing processes. The raw materials used are: (1) elastomers, (2) carbon black, (3) plasticizers, (4) pigments, (5) cord and fabric activators, (6) plastic resins, and (7) other chemicals.

Solid wastes are generated by almost all manufacturing steps. Waste quantities generated in the tire, rubber and plastic footwear, rubber reclamation, rubber hose and belt, and fabricated rubber goods industry segments are listed in Table 25.[28] This table shows that some solid wastes in the rubber products industry originate from spills, air

Table 22
METAL CONCENTRATIONS IN VARIOUS REFINERY SLUDGES[19]

Metals	Dissolved air flotation floats	API Separator sludge	Wastewater treatment plant sludges	Spent filter clays	Crude oil tank bottoms	Distillate and residual petroleum product tank bottoms	Fluid catalytic cracker catalyst fines	Once through cooling water sludge	Cooling tower sludge
Arsenic	0.006—7.7	0.0—310.0	0.0—43.0	0.0—1,100.0	0.003—50.0	0.003—1.0	0.0001—155.0	0.12—5.2	0.01—73.0
Cadmium	0.005—4.0	0.0—100.0	0.0—13.0	0.0—300.0	0.0—10.0	0.005—3.0	0.0001—250.0	0.08—1.6	0.0—22.0
Total chromium	0.3—3000.0	0.022—10,800.0	4.0—1,800.0	0.0—326.0	0.1—83.0	0.03—198.0	0.09—780.0	0.51—350.0	0.3—61,000
Copper	0.08—205.0	0.017—4,000.0	1.3—473.0	0.25—1,000.0	0.07—500.0	0.106—45.0	0.001—600.0	2.17—1,200.0	0.02—16,050
Lead	0.0—540.5	0.0—6,200.0	0.0—1,700.0	0.3—410.0	0.1—250.0	0.13—1,340	1.0—750.0	15.2—280.0	0.03—5,600
Mercury	0.013—0.94	0.003—83.0	0.006—6.2	0.002—6.0	0.0—26.3	0.0003—0.98	0.0—132.5	0.05—21.0	0.0—4.0
Nickel	0.04—68.4	0.006—1,030.0	0.2—400.0	0.3—319.0	0.46—62.6	0.09—130.5	6.3—3,365.0	0.85—300.0	0.0—300.0
Selenium	0.007—4.0	0.0—282.0	0.01—63.0	0.0—118.0	0.0—15.0	0.0007-1.0	0.0—130.0	0.03—10.0	0.0—81.0
Vanadium	0.04—146.9	0.0001—786.0	0.01—50.0	0.0—450.0	0.54—223.4	0.001—190.0	0.01—7,203.0	1.15—23.0	0.31—119.0
Zinc	0.01—4000.0	0.25—9,100.0	0.04—17,000.0	3.4—6,600	0.1—200.0	0.09—538.0	0.5—500.0	2.9—800	0.0080—51,000

Note: Values of concentrations given in mg/kg dry weight.

[a] Solids dried to constant weight at 105°C.

Table 23
CHEMICAL ANALYSIS OF REREFINING WASTES[25]

	Settled sludges		Wastewater		Spent clay		Distillation bottoms	
	Min	Max	Min	Max	Min	Max	Min	Max
Metals								
Arsenic	0.03	22	0.013	24	<0.02	24	<0.01	15
Barium	0	300	0.21	3,610	<1	628	6	1,400
Cadmium	0	37	0.02	216	0.5	13	0	29
Chromium	0	68	<0.5	2,130	5	169	7	100
Lead	<0.1	2,300	0.02	7,770	<1	1,200	1,090	15,000
Zinc	<0.005	1,650	0.09	3,150	52	800	85	3,500
Chlorinated solvents								
1,1,1-Trichloroethane	12	1,900	19	110,000				
Trichloroethylene	20	2,600	2.2	1,300				
Tetrachloroethylene	3.3	1,300	70	8,200				
Total chlorine	76	4,170	88	181,000				
Other organics								
Benzene	<0.4	890	—	—				
Toluene	14	5,800	—	—				
Benzo(a)anthracene	<0.02	<1	—	—				
Benzo(a)pyrene	<0.02	<1	<1	12				
Naphthalene	0.7	700	—	—				
PCBs	0.04	14	<1	500				

Note: Values given are concentration in ppm.

pollution control systems, and rejects. The floor sweepings and dust are usually fine powders containing the raw materials used in the particular manufacturing process.

B. Tire and Inner Tube Industry Manufacturing Process

The tire and inner tube industry represents 66% of the rubber products industry.[28] The main products from this subgroup are pneumatic tires for passenger cars, motorcycles, trucks, buses, and off-highway vehicles. The following raw materials are used in the manufacture of pneumatic tires: elastomers (57.5%), carbon black (20.6%), cord and fabric (7.8%), pigments (6.1%), plasticizers (4.1%), other chemicals (3.0%), and plastic resins (0.9%).[28] A manufacturing process for passenger tires is illustrated in Figure 9.

The first step in tire manufacturing is the receiving and storage of raw materials. At this step, the only source of solid waste generation is the accidental spillage and leakage of raw materials. This is followed by compounding where the raw materials are measured in the proper proportions. Most of the wastes generated at this step are in the form of dust. The nonrubber compounds used are powders which easily become airborne. The particles are captured in baghouses and disposed of as waste.

The compounds are then mixed in a Banbury® mixer.[28] This mixer is a closed container where temperature, pressure, and mixing rate can be controlled. The hot rubber stock is formed into sheets and cooled. This operation generates three types of wastes: (1) dusts from the initial mixing of the compounds, (2) oils from the seals and shafts of the mixer, and (3) "scorched" stock (stock which due to mixing errors becomes too hard for use and has to be discarded).

Each component of the tire is then manufactured independently from this rubber stock and then assembled in the tire building process. The components include white sidewalls, treads, cords and belts, and beads. The manufacture of the components

Table 24
ANNUAL QUANTITY OF SOLID WASTES GENERATED BY STATE IN THE RUBBER AND MISCELLANEOUS PLASTICS PRODUCTS MANUFACTURING INDUSTRY[2-12]

State	1973	1974	1975	1976	1977	1978	1979	1980	1981	1982	1983
Arizona	—	1.1	1.1	13.5	—	—	—	—	—	—	0.0
California	79.0	155.0	160.0	136.0	89.0	128.0	144.0	133.0	125.0	78.0	65.5
Connecticut	—	49.0	24.1	27.0	16.3	—	—	—	—	—	0.0
Florida	2.6	8.0	3.6	4.9	9.0	—	—	—	—	—	0.0
Georgia	—	32.8	30.9	23.5	35.1	—	—	—	—	—	10.4
Illinois	53.5	67.7	39.2	51.6	63.8	86.6	86.5	69.5	71.8	41.6	61.5
Indiana	62.0	70.2	75.1	—	—	—	145.0	122.8	—	69.6	86.4
Iowa	—	—	—	—	—	—	54.4	—	—	46.5	35.8
Kentucky	—	—	—	—	—	—	17.0	—	—	12.9	23.9
Massachusetts	29.8	51.4	38.3	45.5	53.3	—	89.8	—	—	22.9	34.8
Michigan	53.5	63.3	47.9	77.2	50.2	—	—	—	—	23.6	59.4
Minnesota	5.6	7.2	13.6	16.0	30.0	—	—	—	—	—	0.0
Missouri	11.8	39.7	26.8	34.2	8.6	—	—	—	—	—	21.7
New Hampshire	3.8	3.5	3.6	2.2	4.2	—	—	—	—	18.1	21.6
New Jersey	—	—	41.5	68.3	45.7	—	—	—	—	59.0	86.3
New York	18.0	46.4	17.8	39.6	30.2	—	51.8	—	—	26.8	31.7
North Carolina	33.3	38.8	35.3	45.4	56.1	—	—	—	—	60.6	57.6
Ohio	380.9	448.2	316.7	329.3	228.7	311.5	284.7	241.0	224.0	171.9	208.1
Oregon	2.5	7.8	0.5	0.2	—	—	—	—	—	—	0.0
Pennsylvania	69.5	96.5	67.6	86.9	107.6	101.4	114.4	105.9	97.2	57.8	85.4
South Dakota	—	3.6	6.6	7.0	3.2	—	—	—	—	—	0.0
Tennessee	56.6	46.2	44.4	47.8	13.3	—	61.0	—	—	61.2	90.7
Texas	45.9	32.7	34.5	48.8	59.8	—	—	—	—	41.1	55.2
Virginia	—	—	61.0	—	—	—	—	—	—	38.1	37.4
Washington	—	7.4	2.8	5.3	—	—	—	—	—	—	0.0
West Virginia	10.6	17.7	13.0	10.0	1.0	—	—	—	—	—	0.0
Wisconsin	22.2	20.4	14.1	21.7	128.0	—	—	—	—	20.8	21.9

Note: Values given in thousands of short tons.

consists of heating the rubber stock and forming the desired product. For example, in white sidewall manufacturing a white rubber stock is extruded into a black rubber envelope or laminated onto a black rubber backing. The component manufacturing processes generate the following wastes: (1) uncured rubber stock, (2) lubricating waste oils, (3) rejected poor quality or contaminated rubber stock, and (4) off-spec calendered stock. Bead manufacturing involves coating metallic rings with rubber which produces some metal scrap.

Once the compounds are manufactured they are assembled on a rotating drum. First, the inner liner is placed down. Next, a series of cords are placed overlapping the beads, the belts are applied, and finally the tread is wrapped around the tire. This process generates the following solid waste: scrap rubber, reject tires, and end cuts of calendered stock.

The next step is tire painting where the "green" tire is sprayed with agents which release the air from the tire during curing and molding. One possible spray is silicone oil. The tire is then molded and cured into a donut shape. This is done by inflating a bladder bag inside the tire, enclosing a molding around the tire, and applying heat. When the curing is complete, the tire is filled with air and allowed to cool. The main waste generated from this operation is worn-out rubber bladders.

The tire is then finished. Protrusions from the surface are shaved off and the black

Table 25
WASTE TYPES AND QUANTITIES PRODUCED IN RUBBER-RELATED MANUFACTURING IN TIRE AND INNER TUBE MANUFACTURING INDUSTRY[28]

Waste types	Quantity (kg waste per 1000 kg product)
Rubber tire manufacturing	
Floor sweepings containing carbon blacks, pigments, accelerators, antioxidants, etc.	1.9
Dusts containing carbon blacks, pigments, accelerators, antioxidants, etc.	8.3
Scorched rubber	11.3
Uncured stock	3.4
Coated and uncoated fabric	15.9
Green tires	2.7
Rubber, steel wire, and rubber coated fabric	0.7
Rubber dust	4.8
Scrapped tires	21.9
Rubber and plastic footwear	
Floor sweepings and dust from particulate emissions control	0.5
Scrap uncured rubber stock	2.3
Scrap fabric and calendered stock	23
Cut up shoes	4.6
Rubber reclaiming industry	
Dust and floor sweepings	0.1
Unsuitable tires	7.5
Wire scrap	7.5
Fiber and rubber dust	0.3
Separated fiber	11.5
Fine ground fibers	0.4
Oils and contaminated devulcanized agents	1.2
Strainer cleanings	0.3
Rubber and plastics hose and belting industry	
Floor sweepings, dust collected	8.3
Scrap rubber stock	22
Cord rejects	2.8
Cured miscellaneous ends	80
Fabricated rubber goods industry	
Floor sweepings, dust collections	16
Scrap rubber stock	42
Flash trimmings	114
Grindings	23
Rejects	69
Cord rejects	4

casing of the white sidewall is removed. These produce scrap rubber and shavings. Finally, the tires are inspected, and rejected tires are disposed of.

C. Disposal and Reuse Practices

Kushnir and Nagy[28] estimate that 71.4 kg of wastes are produced for every 1000 kg of tire manufactured (Table 25). However, they also point out that waste production in the pneumatic tire manufacturing industry is not directly a function of a plant's tire output because a large percentage of wastes result from accidental spills, material loss during transfer operations, and rejects as off-grade products. These wastes are primarily disposed of in landfills. There is little reuse of these solid wastes.

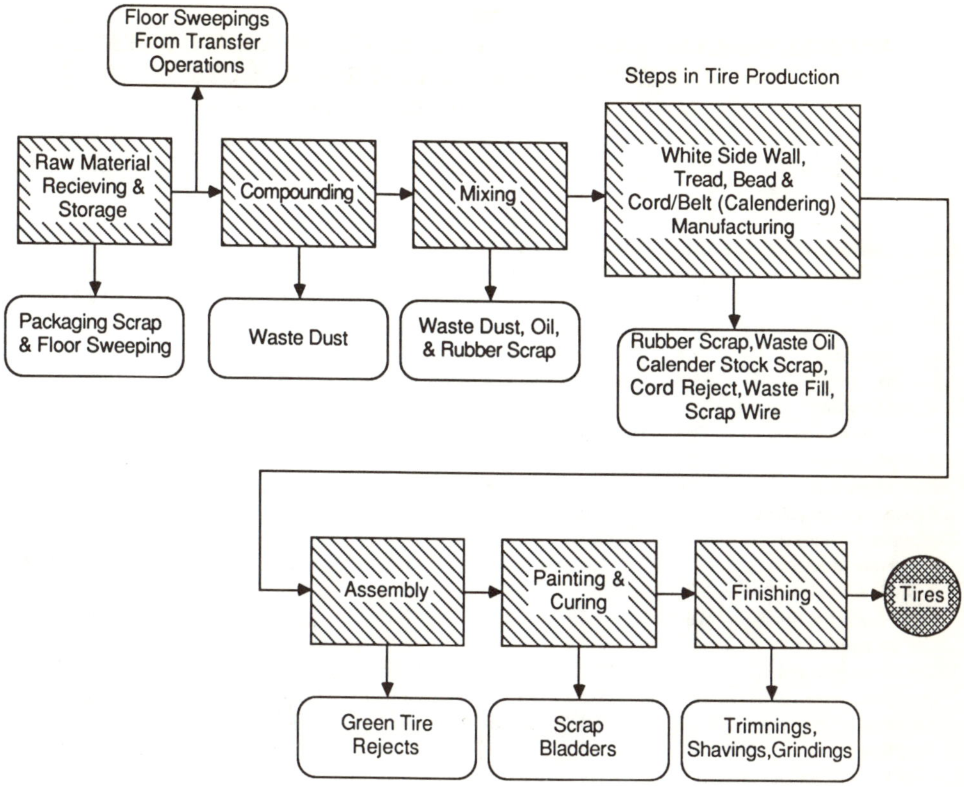

FIGURE 9. Schematic diagram of passenger tire manufacturing.[28]

X. LEATHER AND LEATHER PRODUCTS (SIC CODE 31)

A. Waste Generation

The leather and leather products industry is comprised of installations for tanning, curing, and finishing animal skins into leather. This manufacturing industry is a relatively small generator of solid waste. Table 26 lists data on the generation of solid wastes by state and by year.[2-12] In 1983 this industry generated less than 1% of the total manufacturing of solid wastes. Solid waste generation has averaged 0.3 million tons per year over the last decade. In 1983 only California, Kentucky, New Jersey, Pennsylvania, and Tennessee were manufacturing leather products. The greatest amount of solid waste was generated in New York.

B. Manufacturing Processes

The leather manufacturing process can be divided into the following four steps: (1) beamhouse, (2) tanhouse, (3) retan, color, and fat-liquor, and (4) finishing. Chemicals used in this process include lime, sodium sulfide, sodium sulfhydrate, basic chromium sulfate, vegetable compounds, mineral acids and sodium chloride.[29]

In the beamhouse the hair is removed from the animal hides. This is either accomplished by chemically dissolving the hair with calcium hydroxide, sodium sulfhydrate, and sodium sulfide, or chemically loosening the hair and removing it mechanically.

In the tanhouse the hides are chemically treated so that the hides will not undergo physical or chemical degradation. The most common technique is chrome tanning, which uses chromium, sodium bicarbonate, and formic acid to treat the leather. This

Table 26
ANNUAL QUANTITY OF SOLID WASTES GENERATED BY STATE IN THE LEATHER AND LEATHER PRODUCTS MANUFACTURING INDUSTRY[2-12]

State	1973	1974	1975	1976	1977	1982	1983
Arkansas	8.3	8.4	6.5	4.1	4.8	—	0.0
California	23.0	19.0	18.0	5.0	14.0	9.0	9.4
Delaware	0.9	2.0	0.9	—	—	—	0.0
Florida	1.2	—	1.7	—	1.8	—	0.0
Georgia	—	0.5	—	1.0	—	—	0.0
Kentucky	0.6	1.6	2.7	3.1	3.0	—	0.0
Maine	15.3	10.5	5.5	124.1	37.0	26.5	26.5
Maryland	—	9.3	—	0.2	—	—	0.0
Massachusetts	13.3	12.6	20.2	23.5	60.0	—	0.0
Mississippi	0.3	0.7	—	0.4	0.5	—	0.0
Missouri	17.0	21.0	16.9	13.1	11.8	—	0.0
New Hampshire	1.7	9.0	6.5	3.8	5.8	—	0.0
New Jersey	—	12.7	1.5	6.7	9.0	12.3	20.0
New Mexico	1.7	1.4	—	1.5	—	—	0.0
New York	39.8	—	24.3	—	—	29.8	65.1
North Carolina	18.7	6.3	2.8	2.1	0.8	—	0.0
Oregon	0.4	0.4	7.1	6.9	—	—	0.0
Pennsylvania	—	18.1	—	53.0	14.4	—	3.5
Tennessee	17.5	17.5	17.1	13.5	279.8	13.0	13.3
Texas	0.4	0.4	3.6	—	—	—	0.0
Virginia	—	5.2	—	3.1	1.8	—	0.0
West Virginia	17.4	7.0	5.8	20.3	12.0	—	0.0

Note: Data for years 1978 through 1981 were not available. Values given in thousands of short tons.

process produces high-chromium concentrations in the wastewater treatment sludges. Nonchrome tanning uses either plant tanins extracted from tree bark or synthetic tanins.

Retan is a process where the hides undergo further tanning to obtain additional traits in the leather. This is done with either the original chrome or nonchrome solution with the possible addition of zirconium or glutaraldehyde. The hides are then bleached with sodium bicarbonate and sulfuric acid. Dyes are added to the retanning vats and the leather is colored. In fat-liquoring, the oils removed in the beamhouse processes and in tanning are replaced.

The final step in finishing. This includes smoothing, stretching, drying, conditioning, tumbling, buffing, coating, finishing, and plating. The actual processes used vary from one tannery to another; e.g., when manufacturing leather from pigskins the beamhouse step is not included.[29]

C. Disposal and Reuse Practices

The leather and leather products industry uses large amounts of water and therefore generates large volumes of wastewater treatment sludges. Sludges from tanneries using beamhouse processes and chrome tanning tend to have relatively high chromium, lead, sulfide, and sulfate concentrations; whereas, sludges from tanneries which use chrome tanning but not beamhouse processes have high chromium and lead concentrations. Sludges from tanneries with nonchrome tanning have high concentrations of sulfide and sulfates.[29] Table 27 summarizes metal concentrations of tanning sludges. Chromium concentrations range between 2,000 to 15,000 mg/ℓ in tanning sludges. Copper concentrations are between 100 to 200 mg/ℓ.[29]

Table 27
TYPICAL CHARACTERISTICS FOR TANNERY TREATMENT SLUDGES AND SEWER SUMP SLUDGES[29]

Constituent	Chrome leather tannery pretreatment/treatment sludge: Before dewatering	Chrome leather tannery pretreatment/treatment sludge: After dewatering	Chrome leather tannery sewer sump sludge (not dewatered)	Nonchrome leather tannery secondary treatment sludge (not dewatered)
Solids content %	5—10	20—30	5—15	3—6
Chromium (mg/ℓ)	3,000—6,000	10,000—15,000	2,000—4,000	<5
Copper (mg/ℓ)	100—150	150—200	100—200	<10
Lead (mg/ℓ)	10—25	50—150	10—25	<5
Sulfide (mg/ℓ)	20—50	50—150	30—60	25—50

Table 28
SELECTED CHROME TANNERY WASTE STREAM CONSTITUENTS[29]

Waste stream	Constituent	Mean concentration (wet wt mg/kg)	Concentration range (wet wt mg/kg)
Trimmings and shavings			
Chrome	Chromium	7,600	2,200—21,000
Unfinished leather trim	Chromium	16,900	4,600—37,000
	Copper	90	2.3—468
	Lead	120	2.5—476
	Zinc	60	9.1—156
Finished leather trim	Chromium	14,800	1,600—41,000
	Lead	1,000	100—3,300
Buffing dust	Chromium	5,700	19—22,000
	Copper	960	29—1,900
	Lead	150	2—924
	Zinc	160	—
Finishing residues	Chromium	3,500	0.45—12,000
	Copper	40	0.35—208
	Lead	8,400	2.5—69,200
	Zinc	150	14—876
Sewer screenings	Chromium	2,200	0.27—14,000
	Lead	30	2—110
	Zinc	60	35—128
Wastewater treatment residues (sludges)	Chromium	3,700	0.33—19,400
	Copper	370	0.12—8,400
	Lead	60	0.75—240
	Zinc	50	1.2—147

Other solid wastes generated by the leather manufacturing industries include chrome trimmings, chrome shavings, buffing dust, and sewer screenings. Relatively high concentrations of chromium have been found in these wastes. Buffing dust and sewer screenings also have high concentrations of lead. Table 28 summarizes the metals concentrations of various leather manufacturing solid wastes. Highest chromium concentrations (14,800 to 16,900 mg/ℓ) were present in the leather trim residues.[29] Lead concentrations were highest (8,400 mg/ℓ) in the finishing residues.

Some of this industry's solid wastes can be reused. For example, residual fats can be used in glue, hairs can be used for manufacturing of brushes or in upholstering or rug making, spent tan bark has been used for floor coverings at circuses and horse shows, and chrome tan-liquors are often recycled or reused.[17] Wastes which cannot be reused are landfilled.

XI. STONE, CLAY, AND GLASS PRODUCTS (SIC CODE 32)

A. Waste Generation

The stone, clay and glass group includes flat glass, blown glass, brick and clay tiles, ceramic wall and floor tiles, structural clay products, china, earthenware fixtures, porcelain, pottery, concrete block and brick, other concrete products, ready-mix concrete, lime, gypsum, cut stone and stone products, abrasive products, asbestos products, mineral wool, nonclay refactories and other nonmetal mineral products. Table 29 summarizes the solid waste generation for these industries by state and by year.[2-12] In 1983 the stone, clay, and gravel industries produced roughly 7% of the total manufacturing solid waste. Solid waste generation peaked in 1979 and has since declined. Michigan, California, and Texas are the largest generators of these solid wastes.

B. Manufacturing Processes and Disposal Practices

1. Glass Industry

There are five types of commercially produced glass: soda lime, lead, fused silica, borosilicate, and 96% silica. Soda lime glass is the most common. Raw materials for soda lime glass include silica sand (54.4%), cullet (broken glass 15.4%), soda ash (16.3%), limestone (13.5%), and trace amounts of niter, salt cake, arsenic, and decolorizer.[30] To manufacture soda lime glass these materials are weighed, mixed, and melted. Molten glass is formed, and from this annealed glass, products are manufactured.

Solid wastes include materials collected by air pollution control systems and sludges from wastewater treatment systems. The types and volumes of wastewater sludges depend largely on the process used. The flat glass industry produces an inorganic dense sludge which is typically landfilled. The pressed and blown glass industry produces three types of wastewater sludges: gravity oil separator and dissolved air flotation skimmings, spent diatomaceous earth, and lime precipitation sludges associated with fluoride wastewater treatment. The oily skimmings can be sent to an oil reclamation plant or used directly as road oil. The oily skimmings can also be disposed of by incineration. The diatomaceous earth is typically landfilled. The largest volume of sludge is from the lime precipitation process for fluoride removal. The sludge is vacuum filtered and disposed of in landfills.[30]

2. Clay Manufacturing

Clay manufacturing includes manufacturing of ceramic clay products. In this process the basic ore (kaolinite and montmorillonite) is typically beneficiated and bleached (Figure 10). In the beneficiation process, the raw material is first broken down in a mill. This is followed by screening, selective settling, thickening, and filtering — where the clay minerals are separated from the ore. Finally the clay is dried and bleached. The final product is a white clay which can be used for ceramics manufacturing.

Solid waste is generated primarily in the beneficiation process. Some solid wastes are generated by particulate control systems (e.g., cyclone, wet scrubbers, electrostatic precipitators, and baghouses).[30] These wastes are disposed of in landfills.

3. Stone Manufacturing

Stone manufacturing includes the sand and gravel industry. Sand and gravel is typically mined from open pits. Figure 11 shows a typical sand and gravel processing sys-

Table 29
ANNUAL QUANTITY OF SOLID WASTES GENERATED BY STATE IN THE STONE, CLAY, AND GLASS PRODUCTS MANUFACTURING INDUSTRY[2-12]

State	1973	1974	1975	1976	1977	1978	1979	1980	1981	1982	1983
Alabama	18.4	336.9	4.6	5.5	193.7	—	—	—	—	174.3	12.6
Arizona	—	0.2	0.1	0.1	28.2	—	—	—	—	4.2	16.3
Arkansas	56.9	11.7	6.2	29.1	544.9	—	—	—	—	67.2	0.0
California	426.0	436.0	1339.0	503.0	1347.0	1423.0	1385.0	1417.0	1405.0	328.0	614.0
Colorado	—	8.3	62.2	86.5	194.8	—	—	—	—	—	1.5
Florida	327.8	325.9	370.7	220.1	179.8	—	—	—	—	—	304.5
Georgia	91.7	285.9	80.0	121.3	90.5	93.2	—	109.3	—	118.3	160.8
Hawaii	—	—	32.7	—	0.4	—	—	—	—	—	0.0
Idaho	—	—	—	195.1	10.0	—	—	—	—	—	0.0
Illinois	311.5	301.0	254.1	463.9	426.1	—	—	426.0	—	38.5	53.9
Indiana	402.0	591.5	372.0	—	—	1225.9	—	—	—	272.9	223.4
Iowa	270.9	193.5	320.0	256.5	111.0	—	—	—	—	45.9	171.8
Kansas	—	—	—	—	—	422.9		348.0	357.9	334.5	108.7
Kentucky	161.0	68.1	132.5	100.7	123.3	—	—	—	—	28.6	27.1
Louisiana	175.3	165.3	147.5	140.9	47.5	—	—	—	—	—	0.0
Maine	—	0.3	0.3	0.3	—	—	—	—	—	—	0.0
Maryland	88.7	100.9	67.5	90.3	457.6	—	—	—	—	502.1	147.7
Massachusetts	—	24.1	29.6	20.6	107.7	—	—	—	—	51.8	84.7
Michigan	311.5	811.2	1185.7	1167.8	1598.5	895.0	861.3	781.8	723 .1	569.3	806.1
Minnesota	211.6	254.6	177.5	199.7	32.3	—	—	—	—	156.3	136.7
Mississippi	109.4	216.5	305.5	270.9	210.2	—	—	—	—	—	34.1
Missouri	327.4	1013.0	913.3	285.8	600.0	—	1770.9	1352.7	—	463.2	482.0
Montana	132.1	162.2	168.5	178.1	68.0	—	—	—	—	159.9	151.7
Nebraska	79.6	93.5	178.8	193.9	16.3	—	—	—	—	—	70.0
Nevada	—	—	94.0	230.4	—	—	—	—	—	—	0.0
New Jersey	155.3	189.6	236.2	271.3	614.5	693.7	666.4	620.3	588.0	116.9	104.3
New York	311.8	315.2	388.9	524.0	459.0	501.0	622.9	503.1	511.5	287.8	331.5
North Carolina	52.4	45.7	124.3	132.5	65.4	—	—	—	—	469.9	66.4
North Dakota	34.6	35.3	26.0	35.2	—	—	—	—	—	—	0.0
Ohio	869.7	561.6	562.7	456.4	562.2	720.3	761.7	619.9	505.5	371.6	269.2

Table 29 (continued)
ANNUAL QUANTITY OF SOLID WASTES GENERATED BY STATE IN THE STONE, CLAY, AND GLASS PRODUCTS MANUFACTURING INDUSTRY[2-12]

State	1973	1974	1975	1976	1977	1978	1979	1980	1981	1982	1983
Oklahoma	44.2	46.6	43.4	43.7	25.3	—	—	—	—	—	0.0
Oregon	3.5	2.1	163.7	231.5	15.6	—	—	—	—	—	20.1
Pennsylvania	363.9	556.9	1130.9	1226.5	565.2	—	571.8	569.9	419.1	235.7	198.9
Rhode Island	16.0	19.4	13.1	5.6	3.4	—	—	—	—	—	0.0
South Carolina	—	129.3	109.9	158.5	80.9	—	81.2	93.3	339.1	—	140.3
Tennessee	174.7	75.9	787.1	973.8	1035.5	—	—	—	—	117.8	164.6
Texas	637.4	519.2	329.4	910.5	1471.1	683.0	997.8	1014.1	1485.6	299.9	542.3
Utah	—	—	—	300.4	76.5	—	—	—	—	—	0.0
Virginia	19.0	20.6	47.8	38.4	40.6	—	—	—	—	25.6	19.7
Washington	236.6	104.4	366.4	227.6	517.1	—	—	—	—	5.1	0.0
West Virginia	121.1	451.7	147.5	156.2	44.7	—	—	—	—	208.8	95.7
Wisconsin	25.3	69.0	63.1	67.5	455.1	—	—	—	—	33.1	69.4

Note: Values given in thousands of short tons.

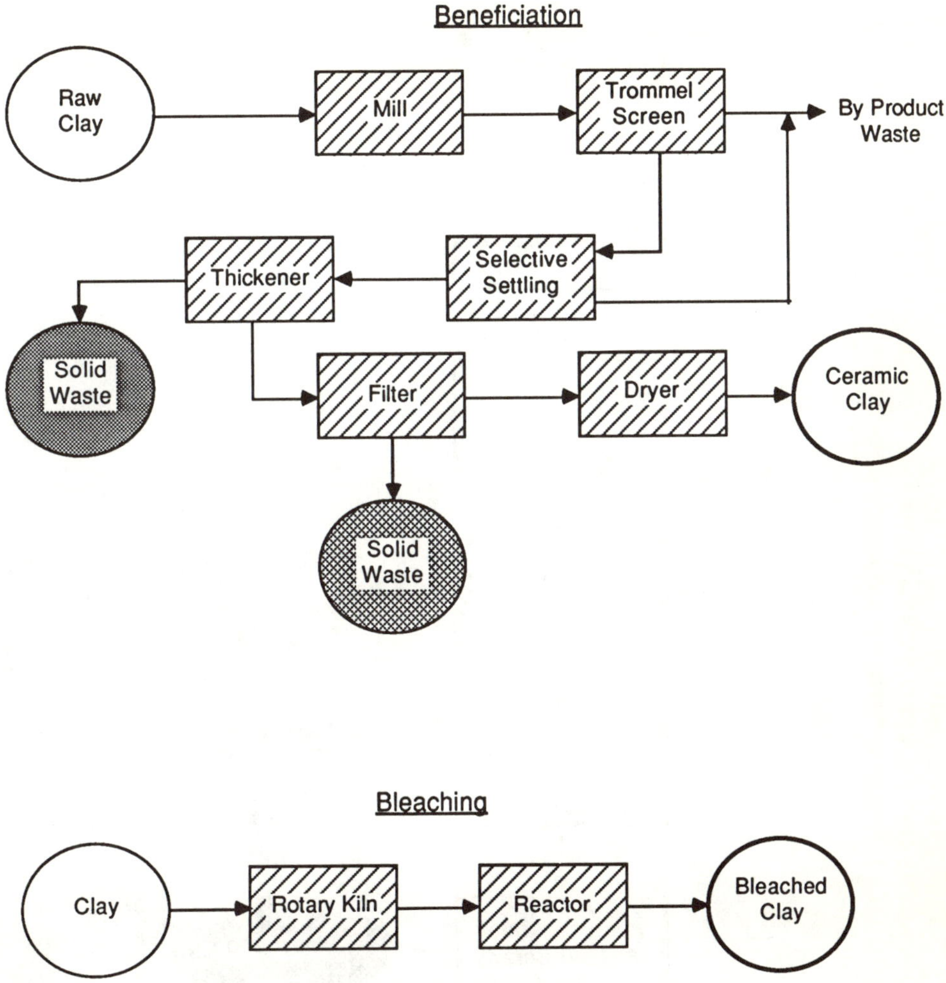

FIGURE 10. Schematic diagram of ceramic clay production.[30]

tem. First, the raw material is passed through separation screens, the rock and gravel is crushed, the sands and gravel then pass through a bar screen separating the gravel and the remaining sands pass through two hydroseparators. In the first separator, the coarse sands are removed; in the second separator, the fine sands are removed.

Between 1 to 20% of the initial raw material remains as solid waste fines.[30] These waste fines are disposed of in the previously mined area, in sedimentation ponds, or on land. If the waste fines are similar to topsoil or dirt, they can be sold. Waste fines that have been mixed with fertilizer are sold as fertilizer or soil conditioner.[30]

XII. PRIMARY METALS INDUSTRY (SIC CODE 33)

A. Overview of Metals Industry

The primary metals industry group consists of facilities such as blast furnaces, steel works, rolling and finishing mills, iron and steel foundries, nonferrous metal foundries, and nonferrous rolling and finishing mills. Foundries are discussed in more detail in Chapter 5. Table 30 lists the annual quantity of total solid waste generated in selected states for the years 1973 to 1983.[2-12] Prior to 1983, Pennsylvania was the leading

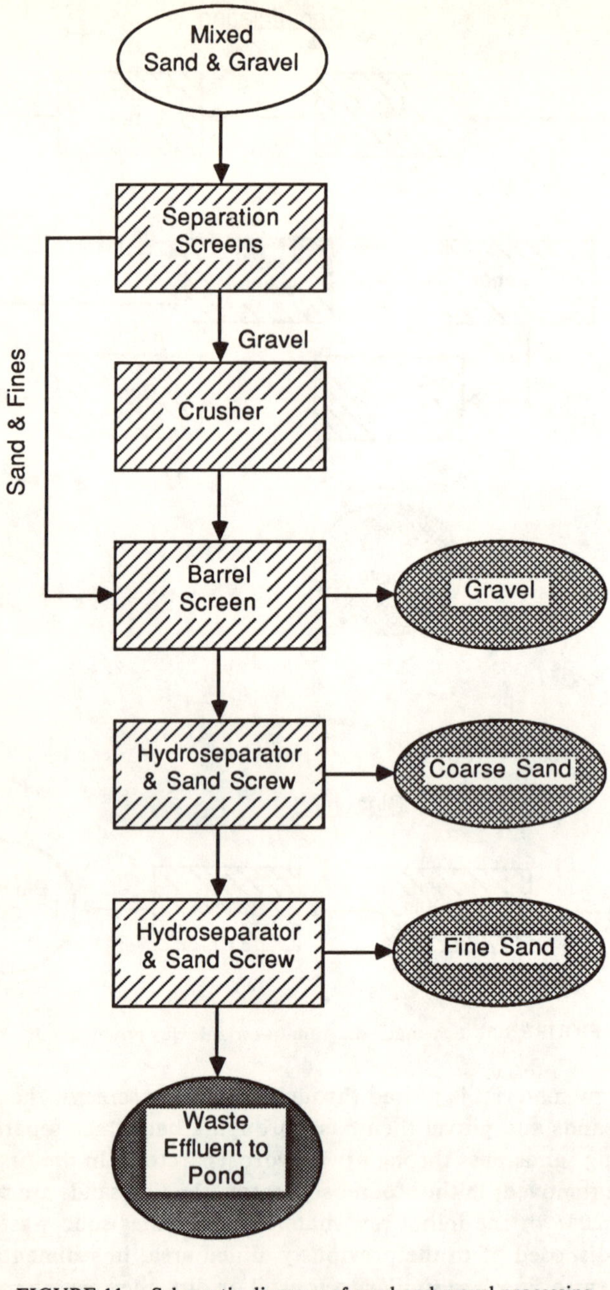

FIGURE 11. Schematic diagram of sand and gravel processing.

metals waste generating state. In 1983, Ohio generated the most waste. Other states which experienced large decreases in the last two years include Arizona (loss of copper smelting), Alabama (loss of steel-making), California, New York, Tennessee, and Texas.

Metals included in the nonferrous group include aluminum, antimony, arsenic, bismuth, cadmium, cobalt, copper, gold, hafnium, lead, magnesium, mercury, platinum, selenium, silver, tellurium, tin, titanium, tungsten, zinc, and zirconium. To illustrate the types of processes used and wastes generated, the manufacturing of iron and steel, and the processing of copper and lead are discussed below.

Table 30
ANNUAL QUANTITY OF SOLID WASTES GENERATED IN SELECTED STATES IN THE PRIMARY METAL INDUSTRIES[2-12]

State	1973	1974	1975	1976	1977	1978	1979	1980	1981	1982	1983
Alabama	178.9	908.2	1215.9	1516.1	2115.5	1587.1	1187.4	1323.1	1181.8	545.5	642.6
Arizona	961.4	4854.4	3736.1	5172.5	4597.2	5454.4	—	115.1	6619.3	86.5	0.0
Arkansas	37.5	47.5	—	—	26.5	—	—	—	—	32.6	0.0
California	1953.0	1843.0	1511.0	1425.0	1476.0	1975.0	1820.0	1609.0	1336.0	110.0	264.4
Connecticut	42.7	60.9	34.9	44.5	62.2	87.5	—	—	85.2	132.5	407.3
Georgia	—	—	—	—	84.2	—	—	—	—	31.5	48.7
Illinois	2229.1	2663.5	2562.9	1480.6	1433.0	1784.6	1520.4	1232.5	1207.6	862.2	1161.7
Indiana	2355.8	3448.6	3264.1	7053.9	5310.3	6278.7	5612.8	3012.6	2270.8	1928.3	1456.6
Kansas	0.3	0.8	0.6	3.2	48.8	—	—	—	—	47.0	33.0
Kentucky	133.4	174.8	221.5	226.7	208.6	584.0	—	—	—	118.0	529.7
Louisiana	—	—	—	—	—	78.7	—	100.4	58.7	—	32.6
Massachusetts	49.4	51.3	86.0	53.8	97.5	—	—	—	—	3.8	44.7
Michigan	2229.1	3130.2	2664.0	3691.1	4282.8	4401.7	4078.2	3155.3	2476.7	1869.8	1283.4
Missouri	103.5	79.6	141.5	136.1	51.6	134.9	291.2	311.6	116.5	142.9	324.3
Nebraska	—	—	—	—	10.9	—	—	—	—	20.5	11.8
New Hampshire	16.4	20.3	3.5	4.2	3.2	—	—	—	—	—	0.0
New Jersey	87.6	61.5	86.4	143.9	160.3	310.1	161.6	146.6	139.1	81.3	297.8
New York	714.6	929.5	1169.0	1180.2	1053.6	1140.1	1524.7	935.6	1058.4	315.1	488.9
North Carolina	7.5	—	—	—	9.8	—	—	—	—	30.9	31.5
Ohio	5993.1	5630.8	5331.5	4020.7	3740.0	4214.7	4211.7	3452.3	3200.4	2245.8	3327.1
Oklahoma	—	—	—	—	—	—	—	—	39.6	160.8	842.1
Oregon	—	113.7	—	—	—	—	180.4	155.8	104.7	50.4	64.2
Pennsylvania	2111.8	9206.9	8284.3	8896.6	8565.1	8321.2	7995.9	5600.1	5717.0	—	1969.7
Rhode Island	30.1	38.8	26.9	24.6	9.3	—	—	—	—	—	0.0
Tennessee	410.0	3415.2	4907.6	88.4	21.0	1613.9	—	1568.3	1480.1	109.0	132.5
Texas	393.2	519.5	796.2	1044.8	1053.3	1069.6	2852.0	896.3	2353.1	1389.1	1215.4
Vermont	0.3	3.0	—	0.6	—	—	—	—	—	—	0.0
Virginia	59.6	246.1	221.4	216.6	165.8	—	201.0	—	—	105.1	164.2
Washington	148.1	197.1	482.1	426.1	386.8	410.5	—	286.9	415.3	475.0	477.8
West Virginia	2165.8	2166.1	2042.9	2050.1	2227.9	2286.5	—	—	—	—	179.6
Wisconsin	240.7	338.7	333.1	479.1	504.3	—	559.6	—	—	384.6	398.2

Note: Values given in thousands of short tons.

B. Ferrous Metal Smelting and Refining Industry (SIC 331-332)

1. Manufacturing Processes

Most of the iron and steel industry is concentrated in a small number of companies (over 92% of U.S. steel is produced by 15 companies). Steel is typically produced by large, integrated mill complexes (Figure 12) which include coke production plants, blast furnaces, steel-making furnaces, continuous casting, ingot molding and roughing operations, hot and cold rolling mills, and facilities for producing special types of steel (e.g., coated or galvanized steel). Coke is produced by heating bituminous coal to drive off volatiles. The coke gas generated by this process is then further processed to recover ammonia, benzene, toluene, and xylene. Wastes generated in a coke plant include ammonia liquor, spent acids and caustics, cooling water, and coke quench water.

Iron ore is converted to molten, elemental iron in blast furnaces to which coke, limestone, and sometimes sintered material has been added. The molten iron is cast as pig iron which is then made into steel in one of three types of furnaces. Open-hearth furnaces use pig iron, scrap iron, lime, and fluorospar to produce steel. Basic oxygen furnaces require less heating time to produce steel than open-hearth furnaces. Electric furnaces can use scrap iron entirely and, in addition, metal alloys may be added to produce stainless steel or ferro-alloys (e.g., ferrochrome or ferrovanadium) instead of low carbon steel. The next step involves molding the steel as ingots or casting it using continuous casting machines. Further processing of the steel into specific products (e.g., sheets, bars, wires, and pipes) may be done at foundries for turning ingots into heavy cast machinery, or at hot or cold rolling mills. Finishing operations involve pickling with sulfuric and hydrochloric acids, annealing and tempering, and coating or plating the steel (e.g., tin-plating by electrolytic deposition or galvanizing by dipping heated steel in molten zinc).

2. Waste Types and Quantities Generated

Large volumes of wastewaters and solid wastes are generated by this industry. The specific waste streams generated by the above processes are also shown in Figure 12.[31] Total solid wastes generated were as high as 28 million tons annually in 1979, but have since declined to about 8.3 million tons in 1983.[2-12] Table 31 provides a breakdown of solid wastes produced by the major component facilities for a typical steel mill complex producing 2.75 million short tons of steel. The largest sources of solid waste on an annual basis are from the blast furnace (613,000 short tons of slag, 28,500 short tons of dust, and 43,000 short tons of sludge) and the basic oxygen furnace (319,000 short tons of slag, 310 short tons of dust, and 37,400 short tons of sludge).[31] In addition to these wastes, sludges are produced from pickle liquor, neutralization solutions, waste ammonia liquor or ammonia still lime sludge, decanter tank tar, mill scales, and large quantities of cooling and rise waters.

3. Chemical Composition of Wastes

The amounts of selected metals and other constituents in steel production wastes are shown in Table 32. With the exception of Cr and Mn, the maximum concentrations of the trace constituents listed (i.e., Cu, CN, F, Ni, Pb, and Zn) are found in the furnace dusts.[32] Maximum concentrations of Cr and Mn are found in electric furnace slag. Phenol is found in decanter tar sludge at levels between 1700 to 3130 ppm. High levels of oil and grease (over 80,000 ppm) are found in the decanter tar sludge and the bar mill scale. Major constituents of slags are shown in Table 33. The composition of the steel slags varies by furnace type.[33] For example, silica, iron, and manganese are the primary components of acid open-hearth slag. Silica, lime, and alumina are the major components of blast furnace slag.

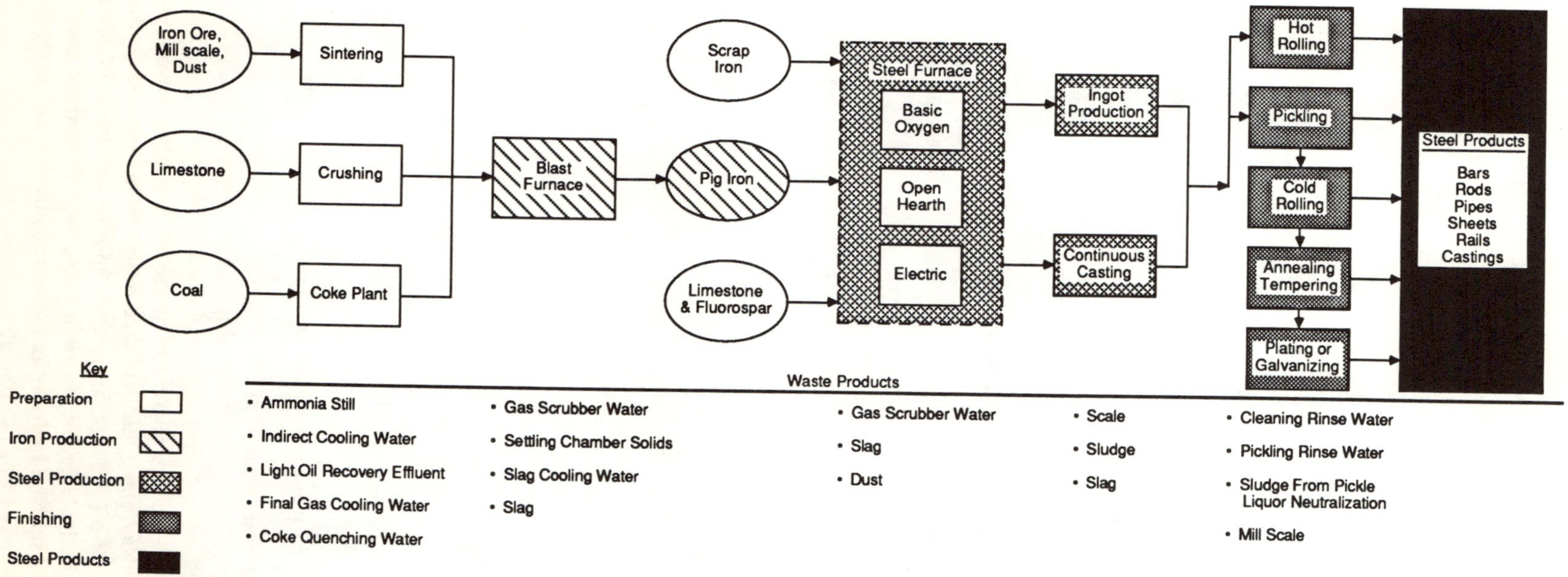

FIGURE 12. Schematic diagram of iron and steel production showing wastes generated.[31]

Table 31
QUANTITIES OF PRODUCTS AND SOLID WASTES GENERATED BY TYPICAL STEEL MILL COMPLEX[31]

Quantities of Products Generated

Facility	Product	Annual amounts
Coke ovens	Coke	1232
Blast furnaces	Iron	1760
Basic oxygen furnaces	Steel	2200
Electric furnaces	Steel	550
Soaking pits	Steel ingots	1720
Primary mills	Billets, blooms, slabs	1480
Continuous caster	Billets, blooms, slabs	870
Hot rolling mills	Sheet steel, bars, rods, structural shapes, etc.	1980
Cold rolling mills	Sheet steel	770
Tin plating mills	Tin plated sheets	110
Galvanizing mills	Zinc coated sheets	140

Solid Wastes Generated

Type of waste	Total quantity of waste
Coke oven — sludge	7
Blast furnace — slag	613
Blast furnace — dust	30
Blast furnace— sludge	43
Basic oxygen furnace — slag	320
Basic oxygen furnace — dust	0.3
Basic oxygen furnace — sludge	38
Electric furnace — slag	66
Electric furnace — dust	7
Electric furnace — sludge	5
Soaking pit — slag	60
Primary mill — sludge	3
Primary mill — scale	67
Continuous caster — sludge	0.1
Continuous caster — scale	7.6
Hot rolling mill — sludge	3.4
Hot rolling mill — scale	36
Cold rolling mill — sludge	0.1
Cold rolling mill — scale	0.04
Cold rolling mill — waste pickle liquor and associated solids	17
Tin plating mill — sludge	0.6
Galvanizing mill — sludge	1.5
Galvanizing mill — waste pickle liquor and associated solids	0.7

Note: Values given in thousands of short tons.

4. Disposal and Reuse Practices

Blast furnace slag is used to produce aggregate for cinder blocks, mineral wool insulation, track ballast, and highway base. Steel furnace slags can be returned to the blast furnace to be used instead of pig iron or scrap steel. Specific quantities of blast furnace and steel slags recycled in the U.S. are shown in Table 34.[33] The largest quantities of slag are used for base and subbase highway and airport construction, aggregate

Table 32
CHEMICAL ANALYSES OF SELECTED CONSTITUENTS IN IRON AND STEEL PRODUCTION SOLID WASTES[32]

Waste type	Cr	Cu	Mn	Ni	Pb	Zn	CN	F	Phenol	Oil and grease	Sn
Basic oxygen furnace ESP dust	270—350	190—220	11,000—12,000	90—140	6,500—8,200	2,600—3,800	—	—	—	—	—
Blast furnace dust	30—170	13—300	2,700—31,000	20—130	50—850	90—1,000	1—40	—	0.1—9.9	—	—
Electric furnace baghouse dust	800—2,700	520—3,400	38,000—45,000	170—500	20,000—25,000	54,000—75,800	—	1,700—3,000	—	—	—
Grinder baghouse dust	20—3,000	400—2,000	1,300—10,100	300—4,400	30—150	20—50	—	—	—	—	—
Grit blast dust	10—1,300	130—2,000	156—8,600	200—800	10—50	40—50	—	—	—	—	—
Open hearth ESP dust	300—1,000	220—3,200	3,000—12,000	190—500	2,700—18,400	850—53,000	—	600—1,400	—	—	—
Reladling baghouse dust	175—200	40—70	2,400—2,800	90—100	40—200	400—980	—	—	—	—	—
Stainless steel grinder dust	37,500	500	2,300	27,000	<10	10	—	—	—	—	—
Basic oxygen furnace slag	40—2,500	10—60	21,000—49,000	<10—50	<10—50	<5—200	—	1,000—8,400	—	—	—
Blast furnace slag	30—70	20—25	2,200—4,100	<10	<10—50	<5—20	—	940—2,200	—	—	—
Electric furnace slag	1,900—8,500	20—120	19,000—81,000	<5—200	<10—200	<5—600	—	500—3,500	—	—	—
Open hearth slag	1,300—5,000	25—70	39,000—61,000	10—50	<10—250	<5—200	—	1,000—4,900	—	—	—
Soaking pit slag	130—600	70—400	3,200—7,300	70—200	<10—2,600	5—150	—	—	—	—	—
Ammonia still lime sludge	40—100	20—35	500—550	5—20	<10—100	550—730	<0.25—43	—	3.4—8.1	19,000—40,000	—
Basic oxygen furnace sludge	220—400	60—115	5,800—11,400	100—800	<10—2,000	250—800	—	740—4,600	—	—	—
Blast furnace sludge	30—150	30—60	1,250—3,000	6—100	450—4,000	1,100—40,000	<0.25—30	680—2,300	0.2—2.4	—	—
Decanter tar sludge	5	1	45	<10	30	20	1—10	—	1,700—3,130	182,000—297,000	—
Electric furnace drum filter cake	760—12,000	580—700	28,000—44,000	200—300	1,800—2,900	3,850—8,900	—	1,700—7,500	—	—	—
Pickle liquor solids	100—5,100	60—1,100	850—15,600	50—750	7—4,500	125—36,700	—	—	—	—	—
Stainless steel processing sludge	23,500	370	560	2,700	<10	50	—	—	—	—	—
Tin plating sludge	30—50	1,100—9,000	190—480	70—700	700—1,100	630—690	—	—	—	—	55,000—67,500
Bar mill scale	190—660	230—350	2,300—4,700	120—270	330—970	25—50	—	—	—	88,100—204,000	—
Hot strip mill scale	20—600	80—800	1,000—2,700	78—6,500	<10—700	10—150	—	—	—	5,000—52,600	—
Plate mill scale	70—100	170—220	4,800—5,700	190—240	<10	9—20	—	—	—	2,700—18,500	—
Slab mill scale	240—500	560—1,400	2,200—7,000	530—1,200	20—200	15—140	—	—	—	2,500—14,000	—

Notes: All units are ppm. Based on data from nine plants sampled up to four times in December 1975 and January 1976. Not all wastes were present at all plants.

Table 33
CHEMICAL COMPOSITION OF STEEL AND BLAST FURNACE SLAG[33]

	Steel slags by furnace type					
	Open Hearth		Basic Oxygen	Electric		Blast furnace slag
Parameter	Acid	Basic		Oxidizing	Reducing	
Silica (SiO_2)	50—60	10—15	12—16	13	22	33—42
Alumina (Al_2O_3)	—	1—3	3—4	4	T	10—16
Lime (CaO)	3—6	50—55	45—50	41	68	36—45
Magnesia (MgO)	—	5—10	8—12	8	7	3—16
Iron (oxide)	20	5—15	25—35	15	T	<2
Manganese (MnO)	15—20	3—5	6—8	13	T	<1
Sulfur (S)	—	0—2	—	T	T	<1
Phosphorus (P_2O_5)	—	—	—	T	—	—

Note: Values given as percentage. T, less than 1 %.

Table 34
QUANTITIES OF IRON AND STEEL SLAG REUSED[33]

	Quantity of slags by type[a]			
	Blast furnace slags			
	Air-cooled	Granulated	Expanded	All steel slags
Aggregate in				
Portland cement concrete construction				
Structures	1,896	—	—	—
Pavements	374	—	—	—
Bituminous construction (all types)	4,539	—	—	563
Highway and airport construction[b]	8,822	—	—	3,704
Manufacture of concrete block	514	23	1,264	—
Highway construction and fill	—	988	—	3,579
Railroad ballast	3,691	—	—	1,327
Manufacture of cement (all types)	—	444	226	—
Mineral wool	704	—	—	—
Roofing slag				
Cover material	262	—	—	—
Granules	132	—	—	—
Sewage trickling filter medium	41	—	—	—
Agricultural slag, liming	6	61	—	108
Other uses	897	141	28	881
Total	21,878	1,657	1,518	10,162

[a] Values given in thousands of short tons.
[b] Other than in portland cement concrete and bituminous construction.

for bituminous paving, and railroad base. Mill scales, fines from the blast furnace, and some steel furnace dust can be processed in sintering plants to produce a material which can be used instead of iron in the blast furnace. In addition, research has been conducted on using electric furnance dust as a source of lead and zinc. Wastes which are not recycled remain as piles (e.g., slag piles) or are disposed of in on-site ponds or landfills.

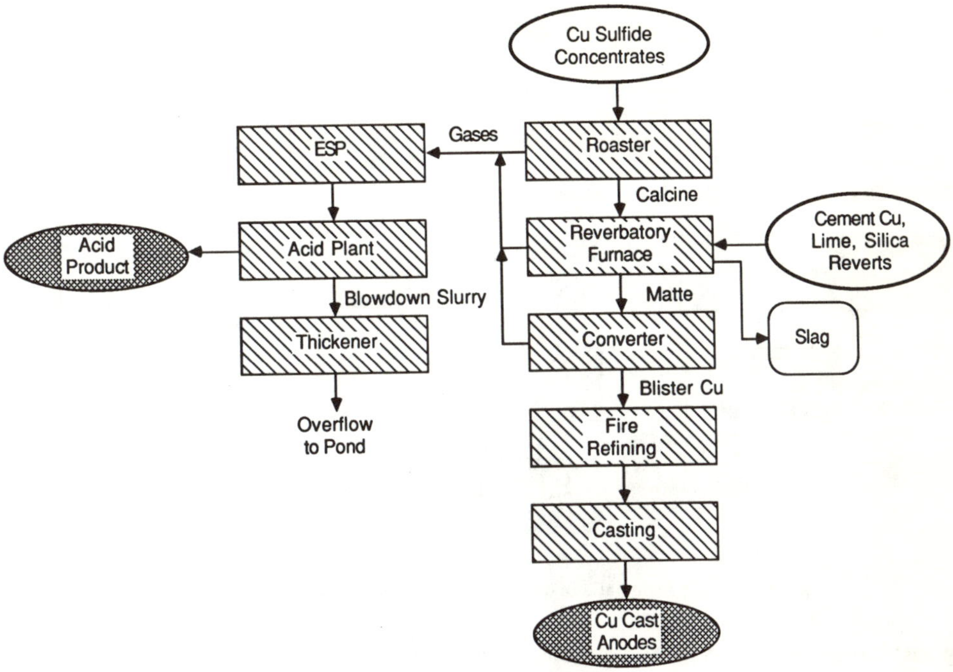

FIGURE 13. Schematic diagram of copper smelting and fire refining process.[34]

C. Copper Smelting and Refining Industry (SIC 3331)

1. Manufacturing Processes

A schematic of the smelting and fire-refining processes is shown in Figure 13. Processing of copper concentrate (typically 25 to 35% copper) involves roasting of the copper to remove volatile elements (e.g., arsenic), smelting to produce a copper and iron sulfide matte (about 45% copper), and converting the ferrous sulfide component into sulfur dioxide and ferrous slag.[34] At this point blister copper has been produced which usually contains a large number of trace metals. The blister copper is further processed using either a fire-refining or electrolytic process. In the fire-refining process, compressed air is brought into contact with the molten metal to oxidize the remaining iron. Sulfur is driven off as SO_2. The iron and other impurities (e.g., tin and zinc) combine with a silica flux to form a slag. Still other impurities (e.g., lead, antimony) are skimmed off as a dross. The copper oxide is next converted to the elemental form (99.7% copper) by heating it in a reducing environment. The copper can be cast directly or made into anodes for electrolytic refining to remove any gold and silver. Electrolytic processing (Figure 14) involves plating of copper from anodes onto thin copper cathodes in an electrolyte solution of copper sulfate and sulfuric acid.[34] Some impurities (e.g., gold, silver, platinum, selenide, and telluride) form an anode slime. Other metals (e.g., arsenic, antimony, bismuth, and nickel) dissolve out of the copper and remain in the electrolyte. Using this process, the copper is 99.95% pure. The copper cathodes are then melted and cast into desired shapes.

In addition to smelting of copper from ore (primary smelting), copper is also recovered from scrap metal (secondary smelting). In 1977, there were 40 secondary copper smelters as opposed to 20 primary smelters. Most of these plants use a pyrometallurgical process rather than an electrolytic refining process. Copper is also recovered from brass mills and foundries. Low grade scrap, slag, and waste drosses are smelted first in a blast furnace to which coke and flux materials (e.g., borax, limestone, and

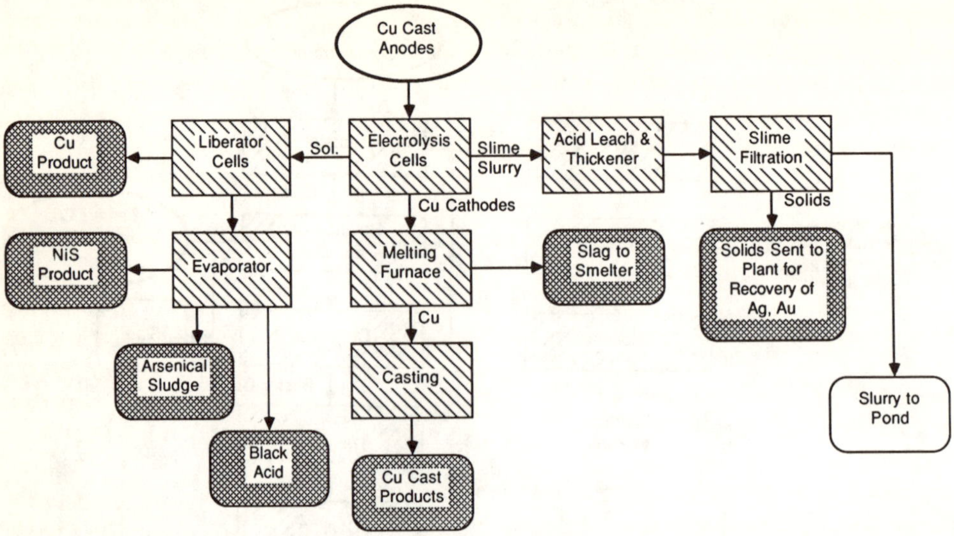

FIGURE 14. Schematic diagram of electrolytic copper refining process.[34]

glass) have been added or in a cupola which operates at lower temperatures. The metal produced is called black copper and contains about 83% copper, 5% zinc, 4% lead, 4% tin, 2.5% iron, and traces of antimony, nickel, and sulfur.[34] The next step is to refine the copper in a converter. If high grade copper is desired, it can then be processed electrolytically. Some of the black copper is used prior to these refining steps to produce brass and bronze alloys.

2. *Waste Generation*

Wastes generated by smelting operations include large quantities of slag, flue dust, sludges, and tailings. For example, at the smelter complex at Anaconda, Montana, which operated between 1902 and 1980, existing residuals include 26.5 million tons of slag covering an area of 130 acres, 38,200 tons of flue dust on a storage pad covering 14.5 acres, and 185 million yd^3 of tailings covering 6000 acres from the Anaconda smelter and Butte concentrator facilities.[35] Other types of wastes generated by smelters include sludges from auxiliary sulfuric acid production plants used to treat SO_2 gases, and waste slurries from the refining, anode production, and casting steps. Electrolytic refining generates anode slime, slime filter cake, arsenical sludge, scrubber slurry, and slags. These slags are up to 50% copper and thus are returned to the smelter. Typical quantities of solid wastes from an electrolytic refinery per short ton of copper produced include 1.2 kg of slime filter cake, 9.8 kg of slag, 2.1 kg of arsenical sludge, and 2.6 kg of solids from miscellaneous slurries.[34]

The types of wastes generated in secondary smelters are similar to those generated in primary smelters. For a hypothetical plant producing 11,000 short tons of copper alloy, the quantities of wastes generated would be 385 kg per short ton of blast furnace slag, reverberatory furnace slag which is recycled to the blast furnace, and baghouse dust which is shipped to zinc smelters for reprocessing.[34]

3. *Physical and Chemical Characteristics*

The physical properties of the waste materials influence the degree to which trace metals can leach out of the wastes. The reverberatory furnace slag consists of black, glassy particles with a median grain size of about 0.83 mm (coarse sand). A typical grain-size distribution is 12% very coarse sand or granules (1.0 mm), 60% coarse sand (1.0 to 0.5 mm), 20% medium sand (0.5 to 0.25 mm), and 8% fine sand (0.25 to 0.125 mm).[35] The average bulk density of the slag is 1.96 g/cm³. Tailings consist of sand-size to clay-size material depending on its source (e.g., jig tailings from a concentrator are coarse sand, while anode slimes are nearly 100% clay). The metals content of the smelting and refining wastes can be high enough to warrant recovery operations. For example, flue dust at one copper smelter averaged 65,300 mg/kg arsenic, 1200 mg/kg cadmium, 85,100 mg/kg copper, 13,000 mg/kg lead, and 20,500 mg/kg zinc.[35] Plans are to reprocess the dust to recover these metals.

Metals content of wastes from six primary and two secondary copper smelters are listed in Table 35.[34] The highest copper concentrations at the plants sampled were in the solids from the acid plant blowdown thickener (up to 380,000 mg/kg) and in the electrostatic precipitator (ESP) dust (240,000 to 280,000 mg/kg). These data show that the lead and zinc concetrations in the acid plant sludge and ESP dust can also be high (greater than 10,000 mg/kg).

Limited leaching test data show that low concetrations of metals can be leached out of the slag, while higher concnetrations can leach out of the flue dust. The metals concentrations from 2 hr column leaching tests[35] on reverberatory furnace slag using acetic acid were about 11 mg/ℓ copper, 0.04 mg/ℓ manganese, 0.02 mg/ℓ lead, and 0.7 mg/ℓ zinc. The pH of the leachate first increased to 7.6 then rapidly dropped to a final pH of 6.3. Sulfate concentrations peaked at approximately 32 mg/ℓ in the first pore volume, then declined to less than 4 mg/ℓ in the second and later pore volumes. Leachate from 7-day column tests of flue dust showed average concentrations of 227 mg/ℓ arsenic, 553 mg/ℓ cadmium, 37,233 mg/ℓ copper, 4.2 mg/ℓ lead, and 23,700 mg/ℓ zinc.[35] Results of column tests on the wastes from the 6 plants listed in Table 35 using 50% solids and water as the leaching solution are listed in Table 36. These tests show a similar pattern of high metal concentrations in the dust samples, low concentrations in the slag, and moderate concentrations in the acid plant sludge.

4. Disposal and Reuse Practices

Smelting and refining wastes are disposed of primarily as piles near the smelter or in tailings ponds. Waste types such as furnace slag and flue dust are isolated from other types of wastes to enhance reuse/recovery operations. Furnace slag is generally placed directly on the land surface. Figure 15 shows an overview of such a slag pile and a tailings pond. In some cases, slags that were disposed of in the tailings pond and have also been used as a base for new tailings ponds. Dusts are stored in closed facilities prior to reprocessing at a smelter, or are stored on lined pads. Because of the high metal content of the dust, the preferred method of handling it is to reprocess it. Tailings from smelter operations are usually discharged to the large tailings ponds used for concentrator wastes. Wastes are typically slurried to the ponds at a solids content of 25 to 30%. Sludges from the smelting and acid plant operations may be disposed of in the tailings ponds or in separate ponds; e.g., at the Anaconda smelter, arsenical sludges were placed in a separate lined impoundment.

D. Lead Smelting and Refining Industry (SIC 3332)

1. Manufacturing Processes

The major processes involved in lead metal production include sintering, charging in a blast furnace, drossing, and further refining to remove impurities (Figure 16). The sintering step involves mixing of lead pellets, silica, sulfur, and iron; heating the materials; oxidizing sulfides to oxides and sulfates, and forming a lead silicate sinter. The sinter is crushed uniformly and fed to the blast furnace along with coke, limestone,

Table 35
TRACE ELEMENT ANALYSES, mg/ℓ OF COPPER SMELTING AND REFINING WASTES[34]

Plant	Type of sample	Cd	Cr	Cu	Hg	Mn	Ni	Pb	Sb	Se	Zn	Sn
Primary copper												
A	Fine dust from ESP on converters	520	50	280,000	0.8	90	110	8,000	500	30	28,000	—
	Reverberatory furnace slag	<5	45	10,000	0.9	230	10	250	250	40	3,700	—
	Fine dust from ESP on reverberatory furnace	310	45	240,000	2.5	100	35	12,000	750	80	44,000	—
	Sludge from lagoon receiving acid plant Blowdown	180	25	22,000	5.0	8	10	>12,000	800	550	1,900	—
B	Electric furnace slag	<5	50	3,700	0.5	165	5	250	<100	10	8,000	—
C	Reverbertory furnace slag	10	100	6,200	0.7	450	25	100	400	20	7,800	—
D	Converter slag	<5	40	40,000	0.5	140	100	200	200	20	1,000	—
	Reverberatory slag	<5	160	6,100	0.8	170	20	80	100	20	650	—
	Solids from acid plant Blowdown thickener Overflow	<5	30	150,000	—	10	<10	10,000	1,200	150	120	—
	Solids from acid plant Blowdown thickener underflow	60	90	380,000	6.0	72	95	5,800	200	40	1,000	—
Secondary copper												
A	Blast furnace slag	<5	20	12,000	—	7,000	260	2,800	<100	—	75,000	—
	Brass casting dross	160	—	340,000	—	—	4,100	—	—	—	330,000	17,000
	Nickel brass dross	160	—	420,000	—	—	1,200	—	—	—	210,000	1,000
B	Chrome brass dross	25	15,500	100,000	—	—	—	—	—	—	460	100,000
	Baghouse dust	5,000	—	47,000	—	—	—	12,000	—	—	520,000	7,500
	Water treatment sludge	10	94,000	170,000	—	—	16,600	900	—	—	1,850	20,000

Notes: — , No data available. Data are from six U.S. plants.

Table 36
RESULTS OF COLUMN LEACHING TESTS[34]

	As	Cd	Cr	Cu	Hg	Mn	Mi	Pb	Sb	Zn	Se	pH	Cond
Primary Copper (SIC 3331)													
Reverberatory slag	0.04	0.01	0.01	0.37	0.02	0.01	0.05	0.3	0.2	0.24	0.09	10.8	600
Electric furnace slag	0.05	0.15	0.02	0.58	0.02	0.2	0.05	6.0	0.2	3.0	0.13	7.8	100
Converter dust	3.44	170	0.9	31,000	0.03	33	15	8.3	2.0	9,000	0.05	3.9	20
Acid plant sludge	0.80	8.4	0.5	850	—	1.0	0.64	7.8	0.2	300	—	3.0	2,700
Reverbatory dust	0.30	130	0.1	29,000	0.01	15	2.5	7.3	0.2	13,000	0.05	4.2	10,000
Secondary Copper (SIC 33412)													
Blast furnace slag	—	1.0	0.03	170	—	0.3	—	6	0.2	55	—	9.4	90
Water treatment sludge	—	0.05	7.1	0.63	—	0.06	—	0.5	0.2	0.01	—	8.6	2,000

Note: — , no data available.

A

B

FIGURE 15. Photographs of (a) copper smelting slag and (b) tailings pile.

and air. In the furnace, the charge separates into slag, matte (44 to 62 weight percent copper), speiss (high in arsenic and antimony), and lead bullion (94 to 98 weight percent lead). The lead bullion is refined by drossing. This process involves first heating the bullion and then cooling it to a temperature which maintains the lead as a liquid at the bottom of the kettle. Copper sulfides and oxides of antimony and other impurities solidify and float to the top. The copper sulfides and oxides are then skimmed off the top and sent to a dross reverberatory furnace for further processing to recover marketable metals by smelting. The slag and molten lead from this operation are returned to the blast furnace. The matte and speiss are shipped to a copper smelter. The lead bullion is further refined by softening which removes the arsenic, antimony, and tin.

2. Waste Generation

The primary wastes generated by lead smelting include blast furnace slag, solids in the slag-granulation slurry, and wastes associated with scrubbing of SO_2 from the

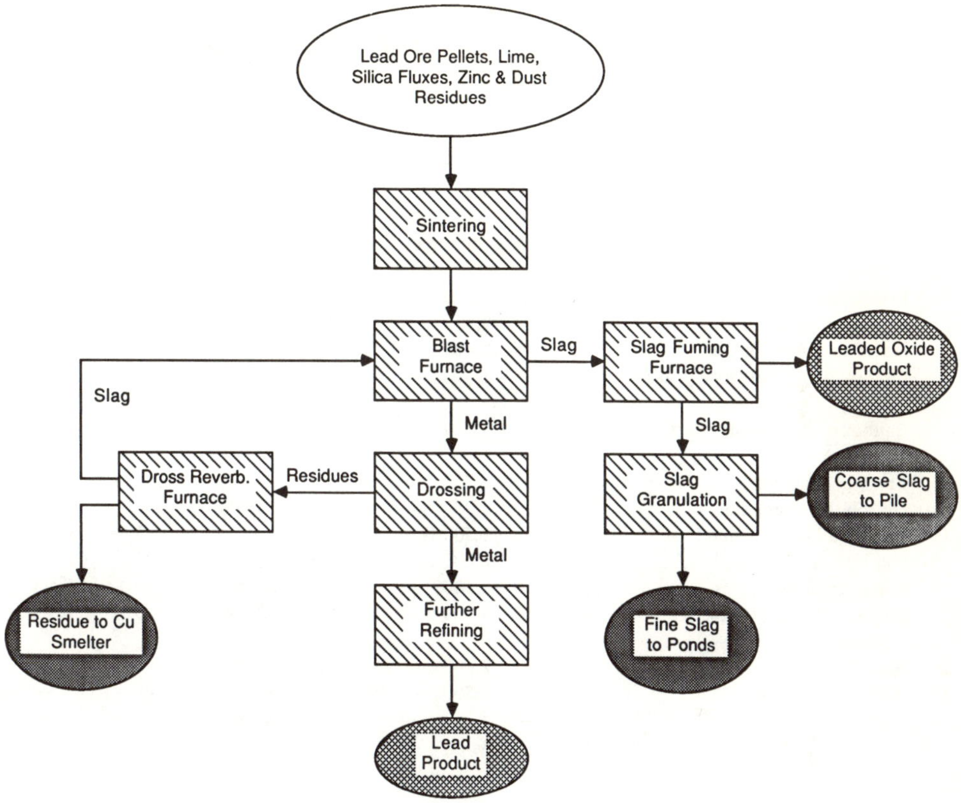

FIGURE 16. Schematic diagram of lead smelting and refining.[34]

gases. The amount of waste generated per year by a typical lead smelter producing 121,000 short tons of lead includes 36,050 short tons of coarse slag, 14,000 short tons of solids dredged from slag-granulating slurry impoundments, and 3850 short tons of fine slag dredged from polishing ponds.[34] Most dust residues produced in the various steps are recycled to the reverberatory furnace.

Wastes generated by the gas scrubbing processes include dust from electrostatic precipitators, wet scrubbers, or baghouses. Depending on the amount of dusts recycled to the sintering plant, the solids sent to ponds would be about 2300 short tons.[34] If the SO_2 gas is used to produce sulfuric acid, acid blowdown slurry and lime treatment sludges are generated. Sludges from an acid plant are about 2400 short tons per year.

3. Chemical Composition of Wastes

Metal concentrations in slags, dusts, and sludges from six primary smelters and three secondary smelters are shown in Table 37. All but one of the slag samples are high in lead. The range was 50 to 164,000 ppm.[32] Zinc, copper, and cadmium concentrations can also be high, particularly in baghouse dust. Leach tests on the lead smelting wastes (Table 38) showed generally low metal concentrations with the exception of copper and zinc in blast furnace dust (130 mg/ℓ and 45 mg/ℓ, respectively).[32]

4. Disposal and Reuse Practices

Most of the slags are dumped close to the smelters but there are some cases of reuse. For example, lead smelter wastes from the Bunker Hill smelter were used to construct part of an embankment for Interstate 90 at Kellogg, Idaho. Approximately 1.2 million

Table 37
CHEMICAL ANALYSIS OF SELECTED METALS IN LEAD SMELTING WASTES[32]

Plant	Waste type	Cd	Cr	Cu	Hg	Pb	Zn	Sb	Tl	Mn	Ni
		Concentration of constituents (ppm)									
		Primary Smelters									
A	Fresh blast furnace slag	10	34	1,850	—	33,500	131,000	—	—	—	—
	Old blast furnace slag	88	37	2,330	—	68,500	51,000	—	—	—	—
	Sinter scrubber sludge	900	11	10,400	0.1	164,000	25,600	—	—	—	—
	Lagoon dredgings (Slag granulation)	700	28	1,490	—	115,000	132,000	—	—	—	—
	Baghouse dust (From blast furnace)	14,000	10	5,350	—	148,000	82,000	—	20	—	—
B	Fresh blast furnace	1,150	37	2,750	—	61,900	110,000	—	—	—	—
	Old blast furnace slag	73	79	2,250	—	46,700	160,000	—	—	—	—
C	Fresh lead fuming slag	10	150	1,500	—	25,000	42,000	33	—	14,560	—
	Old lead fuming slag	5	90	1,600	—	20,000	31,000	20	—	13,500	—
	Blast furnace slag	350	30	1,500	—	94,000	120,000	440	—	11,500	—
	Lagoon dredgings from lead smelter	640	60	6,200	—	140,000	80,000	3,000	—	2,900	—
D	Blast furnace slag	—	—	2,600	—	38,000	108,000	—	—	—	—
E	Blast furnace slag	—	—	2,500	—	35,000	150,000	—	—	—	—
F	Blast furnace slag	—	—	—	—	25,000	120,000	—	—	—	—
		Secondary Smelters									
A	Scrubber sludge	340	30	20	—	53,000	25	1,100	—	120	5
	Wastewater treatment sludge	10	35	80	—	2,700	140	250	—	74	5
B	Smelter furnace slag	5	500	120	—	50	500	100	—	800	5
	Smelter furnace dust[a]	900	150	400	—	120,000	120,000	1,800	—	5	5
C	White Metal Dross[a]	—	—	—	—	160,000	4,700	—	—	—	—
	White Metal Dross[a]	—	—	—	—	145,000	2,600	—	—	—	—

Note: Data are from six U.S. plants denoted A to F. — , not analyzed.

[a] This material is recycled.

yd³ of silty sand tailings were used.[36] Slag-granulating slurry water is discharged to unlined ponds.

Sludges from sulfuric acid production plants and sinter plants are first sent to ponds for dewatering and then are dredged and disposed of in landfills. At some plants, lime treatment sludges are recycled at the plant itself. Matte and speiss from the dross reverberatory furnace can be shipped to a copper smelter to recover the copper.

XIII. FABRICATED METAL PRODUCTS (SIC CODE 34)

A. Waste Generation

The fabricated metal products industries include manufacturers which use raw metal to produce metal products. Examples of processes involved include metal plating, metal coating, can manufacturing, and ornamental metal work manufacturing. There are 36 different product categories included in the fabricated metal products industry, listed by their SIC code in Table 39. Table 40 summarizes the solid waste generated by this industry group by state and by year.[2-12] Solid waste generation has been fairly

Table 38
LEACHATE CONCENTRATIONS OF SELECTED METALS IN LEAD SMELTING WASTES[32]

Waste Type	As	Cd	Cr	Cu	Hg	Mn	Ni	Pb	Sb	Zn	Se
Blast furnace slag	<0.003	0.03	<0.01	0.1	<0.02	<0.01	<0.05	<0.2	<0.2	0.22	0.12
Lead fuming slag	0.089	<0.01	<0.01	1.0	<0.02	0.03	<0.05	<0.2	1.0	0.64	<0.05
Blast furnace dust	0.177	8.0	<0.01	130	<0.02	0.25	0.09	7.3	<0.2	45	<0.05
Sinter scrubber sludge	—	9.1	<0.01	2.6	<0.02	1.3	<0.05	5.5	<0.2	7.5	0.17
Lagoon dredgings (smelter)	0.231	11	<0.01	0.53	<0.02	27	0.08	4.5	<0.2	9.5	<0.05

Note: Values given in mg/ℓ.

Table 39
SPECIFIC PRODUCTS FOR THE FABRICATED METAL PRODUCTS INDUSTRY[37]

SIC	Description
34	Fabricated metal products
3411	Metal cans
3412	Metal drums and pails
3421	Cutlery
3423	Hand and edge tools (except machine tools and hand saws)
3425	Hand saws and saw blades
3429	Hardware, NEC
3431	Metal sanitary ware
3432	Plumbing fixtures and brass goods
3433	Heating equipment (except electrical)
3441	Fabricated structural steel
3442	Metal doors, sash, trim
3443	Fabricated plate work
3444	Steel metal work
3446	Architectural and ornamental metal work
3448	Metal buildings, prefabricated
3449	Metal work, miscellaneous
3451	Screw machine products
3452	Bolts, nuts, screws, rivets and washers
3462	Iron and steel forgings
3463	Forgings, nonferrous
3465	Automotive stampings
3466	Crowns and closures
3469	Metal strappings, NEC
3471	Plating and polishing
3479	Metal coating and allied services
3482	Ammunition, small arms
3483	Ammunition except for small arms, NEC
3484	Weapons, small firearms
3489	Ordnance and accessories, NEC
3493	Steel springs
3494	Valves, pipe fittings
3495	Wire springs
3496	Fabricated wire products, misc
3497	Metal foil, leaf
3498	Fabricated pipe and fittings
3499	Fabricated metal products, NEC

steady for these industries with a maximum in 1974 of 2.1 million tons and a minimum in 1982 of 1.4 million tons. In 1983 the fabricated metal products manufacturers generated roughly 2% of all manufacturing solid waste. Texas, Ohio, Michigan, and Pennsylvania are some of the larger generators of fabricated metal waste.

B. Manufacturing Processes and Disposal and Reuse Practices

As demonstrated by Table 39, the fabricated metals industry includes many types of manufacturing. Typically, the first step in metal fabrication is surface preparation. The purpose of this step is to remove dirt, rust, and residual cutting oils. Five common surface preparation processes are solvent cleaning, alkaline cleaning, acid cleaning, pickling, and etching. Application of solvents dissolves fats, oils, waxes, and other organic materials from the metal surface. Alkaline cleaning removes solids such as grease, oil, and dirt. Acid cleaning removes oxides and minor corrosion from the metal surface. Pickling is the application of a strong acid or base solution to the metal to

Table 40
ANNUAL QUANTITY OF SOLID WASTES GENERATED IN SELECTED STATES IN THE FABRICATED METAL PRODUCTS INDUSTRY[2-12]

State	1973	1974	1975	1976	1977	1978	1979	1980	1981	1982	1983
Alabama	22.3	13.4	14.8	4.9	34.9	—	—	—	—	57.1	15.1
Arizona	0.3	1.7	0.8	2.1	1.0	—	—	—	—	—	0.0
Arkansas	19.5	7.3	13.2	13.4	13.8	—	—	—	—	17.7	0.0
California	84.0	143.0	104.0	121.0	60.0	68.0	74.0	67.0	86.0	51.0	120.8
Colorado	—	13.7	5.3	5.1	8.7	—	—	—	—	—	8.0
Connecticut	40.2	48.8	42.6	—	—	47.8	45.4	41.2	46.3	110.3	50.8
Delaware	0.6	1.0	—	—	0.8	—	—	—	—	—	0.0
Florida	22.0	24.1	14.0	24.9	31.9	—	—	—	—	16.8	19.0
Georgia	18.4	29.4	17.7	19.6	7.7	—	—	—	—	50.5	44.1
Idaho	1.2	1.2	1.4	1.6	—	—	—	—	—	—	0.0
Illinois	220.0	191.8	149.3	155.2	127.2	169.6	189.1	154.1	144.2	71.8	149.5
Indiana	65.9	90.9	54.9	120.0	162.1	144.1	133.2	—	—	82.3	132.1
Iowa	33.6	50.1	66.0	74.4	53.7	—	—	—	—	10.6	32.6
Kansas	—	7.4	—	—	2.2	—	—	—	—	3.9	21.9
Kentucky	25.8	44.8	53.4	50.4	92.1	—	—	—	—	45.4	42.1
Louisiana	10.4	48.7	13.0	12.1	18.0	—	—	—	—	—	3.6
Maine	—	0.5	0.5	0.3	0.6	—	—	—	—	—	0.0
Maryland	11.5	7.2	5.7	6.1	4.9	—	—	—	—	11.3	21.6
Massachusetts	20.3	45.1	20.2	24.6	52.6	—	—	47.5	34.0	14.1	34.1
Michigan	220.0	211.4	127.5	137.2	162.0	163.0	153.2	273.5	236.7	160.0	166.7
Minnesota	18.9	42.6	34.9	29.2	29.5	—	—	—	—	21.3	16.3
Missouri	25.7	23.8	17.9	20.2	24.6	—	—	—	—	24.4	35.0
New Hampshire	1.6	—	0.8	1.8	5.5	—	—	—	—	—	0.0
New Jersey	25.5	47.7	113.0	121.9	59.3	44.4	44.2	33.4	29.7	45.7	63.3
New York	278.4	331.2	453.3	453.7	130.1	59.3	74.9	87.4	37.2	49.7	88.4
North Carolina	4.3	8.5	3.4	4.0	12.2	—	—	—	—	11.6	25.2
North Dakota	10.9	1.1	—	6.5	—	—	—	—	—	—	0.0
Ohio	216.9	208.6	177.8	155.7	201.3	241.0	241.8	248.9	242.5	127.3	171.4
Oklahoma	2.6	2.5	5.0	3.4	4.8	—	—	—	—	3.3	9.9
Oregon	3.0	1.7	1.7	0.6	7.2	—	—	—	—	—	5.7
Pennsylvania	130.4	195.3	159.3	133.5	187.4	136.4	178.8	119.7	146.9	110.1	161.6
Rhode Island	0.6	0.2	0.5	—	4.1	—	—	—	—	—	6.9
South Carolina	—	—	—	—	14.1	—	—	—	—	27.0	18.8
South Dakota	—	0.1	—	—	0.1	—	—	—	—	—	0.0
Tennessee	27.5	35.0	23.5	30.1	39.6	—	—	—	—	11.3	25.2
Texas	31.4	45.6	54.3	53.2	79.9	65.6	100.5	—	60.1	62.2	198.5
Utah	2.3	2.8	—	4.4	—	—	—	—	—	—	0.0
Virginia	5.1	24.9	6.0	3.8	22.7	—	—	—	—	23.3	21.5
Washington	1.0	1.2	1.4	1.6	6.6	—	—	—	—	9.5	3.3
Wisconsin	73.0	147.4	149.6	226.9	239.1	—	—	93.6	96.7	147.9	100.7

Note: Values given in thousands of short tons.

remove scale and rust from the metal surface. Etching involves applications of acid to corrode the surface of the metal prior to further processing.

These surface preparation processes generate liquid waste streams. Solvents are commonly recycled. The acidic or basic waste streams contain trace quantities of dissolved metals. Other solid waste types generated from these processes include sludges from treatment of spent acid, and sludges from treatment of alkaline solutions using neutralization followed by precipitation of metals. The resulting sludges are typically landfilled.

Metal coating (e.g., galvanizing where a coating of zinc is applied to an iron or steel surface to control corrosion) is another common process at fabricated metal plants.

After surface preparation of the metal the zinc is applied either by electrogalvanizing (a type of electroplating), or hot dip galvanizing. This process produces a solid waste called dross, an alloy of iron and zinc which forms at the bottom of galvanizing baths. Metals sometimes can be recovered from the dross or the dross can be landfilled.

Wastes from the various fabricated metal product processes include metals, ceramics, sand, slag, scale, coatings, spent solvents, lubricants, and pickling liquid. Whenever economically feasible, these materials are recovered for reuse. Solid wastes are landfilled.

XIV. ELECTRIC AND ELECTRONIC EQUIPMENT (SIC 36)

The electric and electronic equipment industries include manufacturers of machinery, apparatus, and supplies, which generate, store, transmit, transform, and utilize electrical energy. This group includes:

- SIC 361 Electrical transmission and distribution equipment
- SIC 362 Electrical industrial apparatus
- SIC 363 Household appliances
- SIC 364 Electrical lighting and wiring equipment
- SIC 365 Radio and television receiving equipment
- SIC 366 Communication equipment
- SIC 367 Electronic components and accessories
- SIC 369 Miscellaneous electrical equipment and supplies

A. Waste Generation

Table 41 summarizes the solid waste generated by these industries by state and by year.[2-12] Solid waste generation peaked in 1979, when over 23 million short tons of solid waste were generated. In 1983, the electric and electronic equipment manufacturers generated roughly 2% of all manufacturing solid wastes. Similar to the fabricated metal industry, waste streams are primarily solvents, acids, pickling and cleaning liquors, and rinse waters.[37] On-site treatment of these waste liquids generates various types of sludges. In some cases, pretreatment of waste liquids (e.g., for removal of chromium and cyanide) is conducted on-site prior to discharge of these wastes to municipal sewer systems. Specific solid wastes include metal scrap, carbon glass, metals, rubber, plastics, resins, fibers, and cloth residues.[13]

B. Manufacturing Processes

Electronic components manufacturing (SIC 367) is of particular interest due to the steady growth of the computer industry over the last decade. This subcategory includes many processes. There are nine four-digit SIC codes in this category:

- 3671 Electron tubes, receiving type
- 3672 Cathode ray TV picture tubes
- 3673 Electron tubes, transmitting and special purpose
- 3674 Semi-conductor and related devices
- 3675 Capacitors for electronic applications
- 3676 Resistors for electronic applications
- 3677 Electronic coils and transformers and other inductors
- 3678 Connectors for electronic applications
- 3679 Electronic components, not elsewhere classified

The processes used by these nine industries are diverse and cannot be summarized in a single process flow diagram. As an example, Figure 17 is a process diagram from the

Table 41
ANNUAL QUANTITY OF SOLID WASTES GENERATED IN SELECTED STATES IN THE ELECTRONIC EQUIPMENT MANUFACTURING INDUSTRY[2-12]

State	1973	1974	1975	1976	1977	1978	1979	1980	1981	1982	1983
Alabama	9.3	10.6	9.0	10.4	7.2	—	—	—	—	7.1	9.8
Arizona	—	—	—	—	28.4	—	—	—	—	12.6	14.7
Arkansas	18.3	12.2	7.5	7.7	6.6	—	—	—	—	21.1	0.0
California	59.0	83.0	79.0	93.0	87.0	—	—	116.0	138.0	79.0	139.1
Connecticut	10.3	10.4	9.0	12.8	10.5	—	—	—	—	17.9	68.4
Florida	13.5	14.3	14.8	20.1	18.9	—	—	30.3	—	18.4	24.3
Georgia	27.6	30.9	33.0	29.4	38.7	—	—	—	—	11.7	16.0
Illinois	58.4	94.4	114.2	127.5	113.3	157.5	112.1	—	78.4	66.2	95.5
Indiana	203.9	158.3	164.8	161.9	151.5	149.3	139.3	133.3	126.0	178.0	130.4
Iowa	14.8	15.5	16.7	16.0	18.9	—	—	—	—	17.2	14.6
Kentucky	56.1	77.6	53.4	54.5	105.5	104.2	105.3	85.8	83.0	12.6	31.3
Maryland	34.1	40.6	47.6	16.7	11.5	—	16.1	—	—	9.2	18.4
Massachusetts	32.4	42.9	45.2	50.1	39.0	—	47.1	55.0	41.2	28.4	63.4
Michigan	98.4	27.3	24.9	27.5	25.8	—	—	—	—	11.5	15.1
Minnesota	24.8	10.7	11.8	8.9	6.5	—	—	—	—	7.7	19.5
Mississippi	8.3	10.5	9.9	8.6	14.1	—	—	—	—	25.7	16.2
Missouri	16.7	14.7	15.6	10.7	13.2	—	19.0	—	—	14.3	32.2
New Jersey	73.8	72.7	69.8	50.7	41.9	—	39.7	37.2	35.3	37.6	78.2
New York	80.0	160.8	142.7	141.1	152.0	235.7	717.1	247.1	218.0	130.9	108.0
North Carolina	74.3	24.7	17.7	17.6	33.1	—	27.6	27.1	25.5	24.8	35.8
Ohio	108.3	127.6	110.5	120.4	149.8	298.2	263.0	240.5	208.1	153.0	164.4
Oregon	1.9	3.4	1.6	1.9	1.0	—	—	—	—	—	0.0
Pennsylvania	126.9	172.4	125.9	126.8	218.0	174.6	233.1	192.4	192.5	125.8	169.6
South Carolina	9.0	2.4	3.3	4.9	3.9	—	—	—	—	11.7	21.5
Tennessee	101.9	105.4	46.7	52.8	22.3	—	67.6	—	99.0	45.8	40.6
Texas	31.8	29.8	38.2	72.8	18.5	—	24.8	24.8	33.5	90.4	148.8
Utah	1.5	—	—	—	0.6	—	—	—	—	2.2	3.5
Virginia	27.4	15.2	11.9	17.0	14.3	—	—	—	14.1	9.0	13.3
Wisconsin	29.2	37.2	35.5	36.5	34.8	—	31.3	—	28.8	13.6	56.3

Note: Values given in thousands of short tons.

semiconductor industry for the manufacturing of integrated circuits.[38] First, a silicon ingot is sliced into wafers. The wafers are cleaned with halogenated and nonhalogenated solvents, producing solvent wastes. The silicon wafers are then etched with sulfuric or hydrofluoric acid and rinsed. This produces both a concentrated and dilute acid waste. Next, the wafers are baked, oxidizing the surface. Photoresist, a photosensitive emulsion, is applied. The wafer is then exposed to light through a glass mask that only allows the light to penetrate specific areas. The exposed photoresist becomes hard and sticks to the surface of the silicon, and the unexposed portion is then stripped off with xylene producing a nonhalogenated waste stream. Next, metals (dopants) are diffused into the wafer at selected points by vacuum metalization. These metal impurities act as electrical contact points. Wastes from cleaning the vacuum metalizing equipment generate precious metals which can be recovered, and nonhalogenated solvent wastes. The doped or vacuum metalized wafers are then stripped of the remaining photoresist using H_2SO_4 or H_2O_2, producing concentrated and dilute acid. The wafers are then dried, producing waste halogenated solvents. This is followed by testing and scribing where the wafer is sliced into silicon chips resulting in silicon scrap. Finally, the silicon chips are solvent cleaned, baked, and assembled. The solvent cleaning produces halogenated solvent wastes.

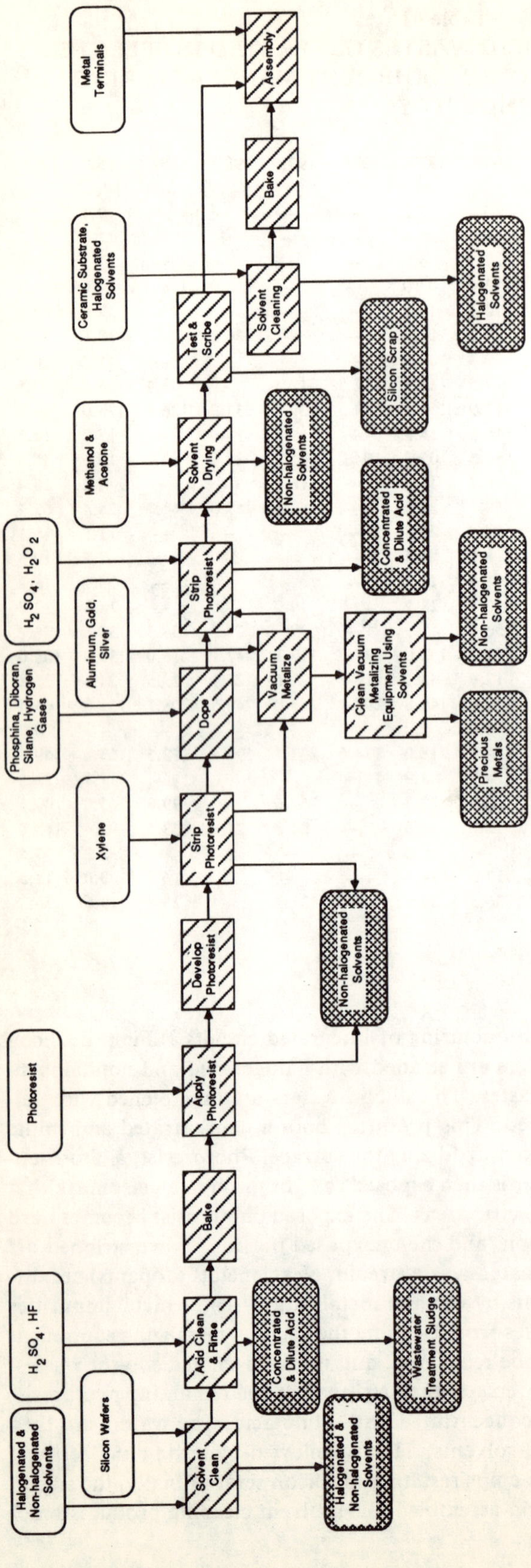

FIGURE 17. Schematic diagram showing steps in integrated circuit manufacturing.[38]

Table 42
CHEMICAL COMPOSITION OF SEMICONDUCTOR WASTEWATERS[38]

Parameter	Min conc (mg/ℓ)	Max conc mg/ℓ	Mean conc mg/ℓ
1,2,4-Trichlorobenzene	<0.01	27.1	4.643
1,1,1-Trichloroethane	<0.01	7.7	1.395
Chloroform	<0.01	0.05	0.015
1,2-Dichlorobenzene	<0.01	186.0	15.972
1,3-Dichlorobenzene	<0.01	14.8	1.450
1,4-Dichlorobenzene	<0.01	14.8	1.341
1,1-Dichloroethylene	<0.01	0.071	0.029
2,4-Dichlorophenol	<0.01	0.017	0.012
Ethylbenzene	<0.01	0.107	0.021
Methylene chloride	<0.01	2.4	0.244
Naphthalene	<0.01	1.504	0.214
2-Nitrophenol	<0.01	0.039	0.024
4-Nitrophenol	<0.01	0.18	0.061
Phenol	0.014	3.5	0.519
Di-*n*-octyl phthalate	<0.01	0.01	0.01
Tetrachloroethylene	<0.01	0.80	0.122
Toluene	<0.01	0.14	0.018
Trichloroethylene	0.007	3.5	0.322
Total Toxic Organics	0.092	245.272	16.248
Antimony	<0.001	0.187	0.021
Arsenic	<0.003	0.067	0.018
Beryllium	<0.001	<0.015	0.002
Cadmium	<0.001	0.008	0.003
Chromium	<0.001	1.150	0.129
Copper	<0.005	2.588	0.570
Cyanide	<0.005	0.01	0.005
Lead	<0.04	1.459	0.145
Mercury	<0.001	0.051	0.004
Nickel	0.005	4.964	0.502
Selenium	<0.002	0.045	0.021
Silver	<0.001	0.013	0.005
Thallium	<0.001	0.012	0.015
Zinc	0.001	0.289	0.093
Phenols	<0.002	6.1	0.630
Oil and grease	ND	20.8	5.058
Total suspended solids	ND	203	31.61
Total organic carbon	ND	80	55.676
Biochemical oxygen demand	9	202	52.768
Fluoride	ND	330	62.0

Note: ND, Note detected. Data are from 11 facilities.

C. Disposal and Reuse Practices

The manufacturing of integrated circuits generates concentrated and dilute acid wastes, concentrated alkaline solutions, spent cyanide solutions, halogenated and non-halogenated solvent wastes, precious metals from vacuum metalization, metal and silicon scrap, wastewater sludges, paint wastes, oils, and plastics. Table 42 presents chemical composition data for wastewater from 11 semiconductor facilities. The waste contained over 10 mg/ℓ of 1,2,4-trichlorobenzene, 1,3-dichlorobenzene, and 1,4-dichlorobenzene; over 1 mg/ℓ of chromium, copper, lead, phenol; and up to 330 mg/ℓ of

Table 43
CHEMICAL ANALYSIS OF WASTEWATER TREATMENT SLUDGES FROM ELECTRONIC COMPONENTS MANUFACTURING PLANTS[38]

Chemical parameters	Sludge from electroplating wastewaters[a]		Metal oxide sludge[a]	
	Extracted metals[b]	Total metals[c]	Extracted metals[b]	Total metals[c]
Cadmium	<0.02	102	<0.06	1.8
Chromium	<0.04	700	<0.04	22
Copper	2.6	1,650	0.4	120
Iron	0.2	7,850	0.4	106,800
Lead	<1.0	58	<1.0	2.7
Zinc	<0.02	105	3.26	38,700
Nickel	0.5	1,980	1.2	1,890
Manganese	—	—	30	40,800
Percent solids (dried at 550°C)	8.40		95.71	
pH	8.1		7.67	

[a] Values given in ppm.
[b] Extracted metals are obtained by partial dissolution with dilute acid solution.
[c] Total metals are measured by strong acid digestion of the waste.

fluoride.[38] Waste rinse waters, plating solutions, and spent acids are typically treated by pH adjustment, flocculation, and precipitation. Separate treatment units are sometimes used to remove fluorides, ammonia, and cyanides prior to discharging the wastewater off-site. Wastewaters are then discharged to municipal treatment plants or to surface waters. Wastewater treatment sludges (Table 43) have high metal concentrations particularly copper, iron, and nickel.[38] These sludges are disposed in landfills. Spent solvents are typically recycled.

REFERENCES

1. United States Office of Management and Budget, Standard Industrial Classification Manual, U.S. Government Printing Office, Washington, D.C., 1972.
2. United States Department of Commerce Bureau of Census, Pollution Abatement Costs and Expenditures 1973, MA-200(73)-2, Washington, D.C., 1976, 112.
3. United States Department of Commerce Bureau of Census, Pollution Abatement Costs and Expenditures 1974, MA-200(74)-1, Washington, D.C., 1976, 86.
4. United States Department of Commerce Bureau of Census, Pollution Abatement Costs and Expenditures 1975, MA-200(75)-1, Washington, D.C., 1977, 80.
5. United States Department of Commerce Bureau of Census, Pollution Abatement Costs and Expenditures 1976, MA-200(76)-2, Washington, D.C., 1977, 90.
6. United States Department of Commerce Bureau of Census, Pollution Abatement Costs and Expenditures 1977, MA-200(77)-2, Washington, D.C., 1979, 84.
7. United States Department of Commerce Bureau of Census, Pollution Abatement Costs and Expenditures 1978, MA-200(78)-2, Washington, D.C., 1980, 24.
8. United States Department of Commerce Bureau of Census, Pollution Abatement Costs and Expenditures 1979, MA-200(79)-2, Washington, D.C., 1981, 47.
9. United States Department of Commerce Bureau of Census, Polltuion Abatement Costs and Expenditures 1980, MA-200(80)-1, Washington, D.C., 1981, 41.
10. United States Department of Commerce Bureau of Census, Pollution Abatement Costs and Expenditures 1981, MA-200(81)-1, Washington, D.C., 1983, 35.

11. United States Department of Commerce Bureau of Census, Pollution Abatement Costs and Expenditures 1982, MA-200(82)-1, Washington, D.C., 1984, 60.
12. United States Department of Commerce Bureau of Census, Pollution Abatement Costs and Expenditures 1983, MA-200(83)-1, Washington, D.C., 1985, 24.
13. Mantell, C. L., *Solid Wastes: Origin, Collection, Processing, and Disposal,* John Wiley & Sons, New York, 1975, 1127.
14. U.S. Department of Commerce Bureau of Economic Analysis Business Statistics, *Supplement to the Survey of Business 1982,* 23rd ed., U.S. Department of Commerce, November 1983(Suppl.), 6.
15. U.S. Department of Commerce Bureau of Economic Analysis, *Survey of Current Business,* Vol. 65, No. 6, 1985, 5-2.
16. Marshall, K. R. and Harper, W. J., The treatment of wastes from the dairy industry, in *Food and Allied Industries, Surveys in Industrial Wastewater Treatment,* Vol. 1, Barnes, D., Forster, C. F., and Hrudley, S. E., Eds., Pitman Publishing, Boston, 1984, 296.
17. Booz-Allen, Applied Research, Inc., A Study of Hazardous Waste Materials, Hazardous Effects and Disposal Methods, Vols. 1, 2, and 3, EPA 670/12-73, EPA-670/14-73, EPA-670/15-73, EPA-670/16-73, U.S. Environmental Protection Agency, Washington, D.C., 1973.
18. Development Document for Effluent Limitations Guidelines and New Source Performance Standards for the Textile Mills Points Source Category, EPA-440/1-74-022a, U.S. Environmental Protection Agency, Washington, D.C., 1974, 241.
19. Brown, K. W. and Associates, Inc., Hazardous Waste Land Treatment, EPA-SW-874, Environmental Protection Agency, Washington, D.C., 1980, 974.
20. Development Document for Effluent Limitations Guidelines and New Source Performance Standards for the Plywood, Hardboard, and Wood Preserving Segment of the Timber Products Processing Point Source Category, EPA-440/1-74-023a, U.S. Environmental Protection Agency, Washington, D.C., 1974, 325.
21. Development Document for Effluent Limitations Guidelines and Standards for the Pulp, Paper, and Paperboard and the Builders' Paper and Board Mills Point Source Categories, EPA-440/1-80-025b, U.S. Environmental Protection Agency, Washigton, D.C., 1980, 633.
22. Kirk, R. E. and Othmer, D. F., Eds., *Encyclopedia of Chemical Technology,* 3rd ed., Vol. 16, John Wiley & Sons, New York, 1978.
23. JRB Associates, A Compilation of Statistics on Solid Waste Management Within the United States, EPA Contract No. 68-d01-6000, U.S. Environmental Protection Agency, Washington, D.C., 1981, 61.
24. Austin, G. T., *Shreve's Chemical Process Industries,* 5th ed., McGraw-Hill, New York, 1984, 859.
25. Franklin Associates, Ltd., Composition and Mangement of Used Oil Generated in the United States, EPA-530-SW-013, U.S. Environmental Protection Agency, Washington, D.C., 1984, 159.
26. Munsen, J. and Vincent, R., *Assessment of Used Solvent and Used Oil as Fuel in California,* State of California Air Resources Board, 1985, 102.
27. Manual on Disposal of Refinery Wastes, First Edition, American Petroleum Institute, Washington, D.C., 1980, 27.
28. Kushnir, J. M. and Nagy, S. F., *Assessment of Industrial Hazardous Waste Practices, Rubber & Plastics Industry,* Foster D. Snell, Inc., EPA-530-SW-163, Environmental Protection Agency, Washington, D.C., 1978, 223.
29. Development Document for Effluent Limitations Guidelines and New Performance Standards for the Leather Tanning and Finishing Point Source Category, EPA-440/1-74-016-a, Environmental Protection Agency, Washington, D. C., 1974, 158.
30. Sittig, M., *Pollution Control in the Asbestos, Cement, Glass and Allied Mineral Industries,* Noyes Data Corporation, Park Ridge, N.J., 1975, 136.
31. Assessment of Industrial Hazardous Waste Practices in the Metal Smelting and Refining Industry, Vol. 3, Ferrous Smelting and Refining, EPA-SW-145c.3, U.S. Environmental Protection Agency, Washington, D.C., 1977, 149.
32. Assessment of Industrial Hazardous Waste Practices in Metal Smelting and Refining, Vol. 4, Appendices, EPA-SW-145c.4, U.S. Environmental Protection Agency, Washington, D.C., 1977, 57.
33. Drake, H.J. and Shelton, J. E., Disposal of iron and steel slag, in *Proc. 4th Mineral Waste Util. Symp.,* IIT Research Institute, Chicago, 1974, 303.
34. Assessment of Industrial Hazardous Waste Practices in the Metal Smelting and Refining Industry, Vol. 2, Primary and Secondary Non-Ferrous Smelting, EPA-SW-145c.2, U.S. Environmental Protection Agency, Washington, D.C., 1977, 308.
35. Tetra Tech, Inc. Lafayette, Calif., Unpublished data, 1985.
36. Pettibone, H. D. and Kealy, C. D., Engineering properties and utilization examples of mine tailings, in 3rd Mineral Waste Utilization Symposium, 1972, Chicago, 161.

37. JRB Associates, Industrial Resource Recovery Practices: Fabricated Metals Production (SIC 34), Machinery Manufacturing/Non-Electrical (SIC 35), Manufacturing of Electrical Machinery (SIC 36), EPA Contract No. 68-01-6000, McClean, Va., 1982.
38. Development Document for Effluent Limitations Guidelines and Standards for the Electrical and Electric Components Point Source Category, EPA-440/1-80/075-a, U.S. Environmental Protection Agency, 1980, 659.

Chapter 3

ELECTRIC UTILITY INDUSTRY SOLID WASTES

I. P. Murarka, R. H. Boyd, and H. P. Harbert

TABLE OF CONTENTS

I. INTRODUCTION

The electric utility industry generates approximately 2300 billion kWh of electricity each year.[1] Production of this vast amount of power is achieved through various technologies using coal, oil, natural gas, hydropower, and nuclear fuels. While all of the fossil and nuclear power plants produce solid wastes, this chapter focuses on solid wastes and wastewaters generated from coal firing. Selection of coal-fired power plants for discussion is appropriate because over 90% of the industry's solid wastes comes from this technology. In fact, the generation of electricity at coal-fired power plants resulted in about 75 million tons of solid wastes in 1983.[2] This chapter presents a discussion of the types of wastes produced, the methods by which they are produced and collected, their characteristics, and methods for treatment, disposal, or reuse.

II. TYPES OF WASTES

By industry convention, and somewhat according to regulatory guidelines, waste materials associated with a coal-fired power plant are divided into two general categories:[3]

1. High-Volume Wastes — fly ash, bottom ash, boiler slag, and flue gas emission control wastes.
2. Low-Volume Wastes — all other wastes not defined as "high volume." These wastes include boiler blowdown wastewater, demineralizer regenerant waste, cooling tower blowdown, metal cleaning solutions, and drains from floors and the station.

Table 1 gives a classification of these wastes to be discussed in this chapter.

III. GENERATION AND COLLECTION OF WASTES

In the U.S. there are over 400 power plants that burn coal as their primary fuel (Figure 1).[4] A complete listing of all operating plants is provided in Appendix A. Because the power plants are quite diverse in design and operation, it is not possible to be all inclusive in our discussion of the wastes from these plants. Therefore, the following discussion is divided along the waste categories of high-volume solid wastes and low-volume wastes. Before we describe the high-volume and low-volume wastes, let us first examine the properties of the fuel coals.

Table 1
WASTE MATERIALS FROM COAL-FIRED POWER PLANTS

Waste type	High volume mostly solids	Low volume Solid	Low volume Liquid
Fly ash	x		
Bottom ash	x		
Boiler slag	x		
Flue gas emission control	x		
Boiler blowdown			x
Demineralizer regenerant			x
Cooling tower blowdown			x
Cooling tower cleaning		x	x
Metal cleaning wastes			
Fireside		x	x
Waterside		x	x
Other equipment			x
Coal pile runoff			x
Pyrites		x	
Yard area runoff			x
Floor and equipment drains			x

A. Properties of Coal

There are three types of coals burned to produce electricity: bituminous, subbituminous, and lignite. The composition and quantity of combustion ash primarily is a function of the composition of these feed coals; therefore, a review of both major and trace constituents contained in coal ash is appropriate. Typically, the ash residue from laboratory combustion of coal has been used to estimate the range of major and trace constituents contained in the various coal types.[5-7] Table 2 gives a summary of the major constituents found in these coal ashes by geographic regions.[8] The major fractions, for all coal ash types, are oxides of silicon, aluminum, iron, and calcium. All coal sources show a large range in these fractions. It should be noted that the western coal ashes are higher in calcium and magnesium oxides, and lower in iron oxide compared to the coal ashes from other geographical regions. The higher levels of calcium and magnesium oxides typically yield an ash which is alkaline in chemical reactions. Lower iron oxide content reduces the potential for iron leaching from disposal of ash.

Table 3 is a summary of the trace constituents found in these ashes by coal types.[9] This table also depicts a large range of values for the various trace elements found in coal ashes.

The amount of ash generated by a coal-fired power plant is in large measure a function of the ash content of the input coal. In general, a high-grade coal, i.e., bituminous, has less ash than a lower grade coal, such as lignite. Proximate analyses of various coals, which include percent ash, volatile matter, fixed carbon, and Btu content have been summarized and are included in Table 4.[10]

B. Generation of High-Volume Wastes

Figures 2 and 3 are simplified schematics of a conventional coal-fired power plant.[11-12] Coal is brought to the furnace, which is surrounded by the boiler tubes, for firing at about 1400°C. The combustion process results in gases and solid products: fly ash, bottom ash, or boiler slag. Collectively, the solid products are referred to as coal ash, or simply ash, and represent the noncombustible fraction of the coal, along with

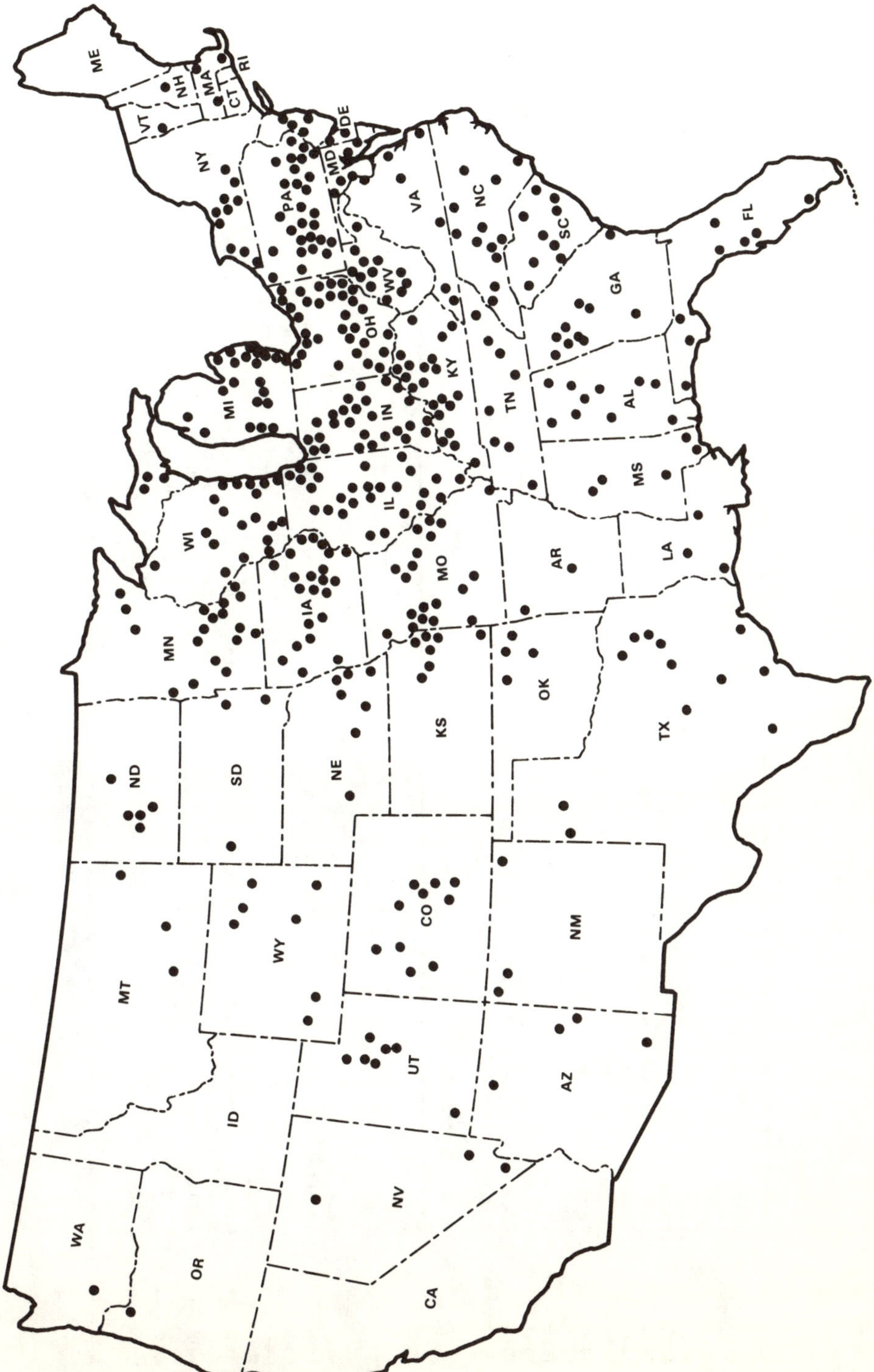

FIGURE 1. Distribution of coal-fired power plants in the U.S.

Table 2
AVAILABLE DATA ON RANGE OF MAJOR CONSTITUENTS IN ASH FROM LABORATORY COMBUSTION OF COALS[8]

	Middle Atlantic (PA)		East North Central			West North Central				South State		
Constituent	PA B(1)	PA A (3)	IL B (1)	IN B (1)	OH B (1)	LA B (1)	KS B (1)	MO B (1)	ND L (2)	MD B (1)	VA B (1)	WV B (1)
SiO_2	26.9—56.5	50.1—58.0	36.0—54.5	30.7—60.9	29.4—56.1	29.0—39.6	35.9—40.5	37.9—45.4	7.7—52.5	46.2—57.1	26.5—64.0	22.0—63.6
Al_2O_3	18.2—32.7	26.5—34.3	14.7—23.2	16.1—31.6	18.1—30.2	12.1—15.8	14.2—18.5	14.5—16.8	2.8—19.8	29.1—31.6	19.1—36.6	14.1—41.6
Fe_2O_3	5.1—52.5	3.51—11.5	16.3—35.4	7.0—14.7	8.6—45.2	32.5—34.3	25.0—40.5	25.8—41.0	2.0—34.3	4.7—16.4	4.5—35.0	2.1—56.3
CaO	12.—9.9	0.25—2.15	1.7—10.4	1.7—8.4	0.4—4.8	4.3—15.0	1.8—11.7	1.7—7.0	7.3—48.7	1.7—2.0	1.0—22.8	0.4—12.7
MgO	0.2—1.4	0.45—1.28	0.4—1.3	0.5—1.5	0.2—1.7	0.9—1.6	0.3—0.08	0.4—0.8	2.8—11.7	0.5—0.8	0.4—4.4	0.2—3.8
Na_2O	0.1—0.5	0.09—0.27	0.1—0.6	0.2—1.1	0.1—0.4	0.2—0.8	0.2—0.4	0.1—0.2	0.1—23.0	0.3—0.9	0.2—1.8	0.1—2.1
K_2O	0.6—3.6	0.38—2.94	1.7—2.6	1.3—3.7	0.4—2.7	1.2	0.4—1.6	1.3—3.0	0.1—1.7	2.5—2.6	0.3—4.6	0.2—4.2
TiO_2	0.7—1.8	1.05—3.35	0.6—1.5	0.8—1.3	0.6—2.2	0.8—0.9	0.6—0.7	0.6—0.8	0.0—2.4	1.2—1.6	0.6—2.1	0.5—3.3
P_2O_5	0.06—2.9	—	0.03—0.27	0.03—0.59	0.04—0.91	0.02—0.56	0.05—0.27	0.02—0.14	0.0—1.4	0.05—0.37	0.04—0.70	0.02—3.0
SO_3	0.2—3.6	0.17—2.55	0.8—2.8	0.2—3.1	0.4—3.6	2.4—3.7	1.4—4.0	1.1—3.5	8.1—33.4	0.4—1.3	0.6—11.2	0.3—9.6
No. samples	31	5	11	13	36		2	4	432	2	26	142

	East South Central			West South Central		Mountain						Pacific
							MT					
	AL B	KY B	TN B	AR B	TX L	CO B	SB	L	NM SB	UT B	WY B	B
SiO_2	23.9—54.0	31.6—57.9	33.6—56.5	24.4—25.2	46.3	34.8—71.8	15.1—59.0	11.2—55.0	21.9—46.2	46.4—63.2	24.5—38.6	37.2—54.1
Al_2O_3	18.4—33.3	18.8—32.9	18.6—32.7	12.1—27.4	17.5	15.2—34.2	8.6—32.1	6.2—31.4	13.8—31.9	9.1—20.3	14.2—19.6	29.7—38.2
Fe_2O_3	5.3—45.0	4.1—31.3	6.1—41.6	20.3—26.5	7.3	3.2—11.9	1.9—27.8	0.9—20.8	2.9—8.0	3.7—19.3	9.0—10.3	2.8—9.2
CaO	1.7—12.4	0.9—5.7	1.7—2.8	7.4—18.8	12.6	0.4—12.8	3.1—45.7	4.5—34.7	1.8—31.4	3.5—21.9	9.4—30.8	1.7—7.6
MgO	0.6—2.4	0.3—2.1	0.7—1.6	4.4—5.4	2.6	0.4—2.9	0.2—11.4	0.8—13.6	1.4—10.4	0.3—7.6	4.4—4.7	0.9—2.6
Na_2O	0.2—0.3	0.2—1.8	0.2—0.5	0.8—2.1	1.1	0.1—3.0	0.1—10.6	0.1—12.0	0.1—8.1	0.4—4.3	0.1—0.2	0.2—1.5
K_2O	0.9—4.0	0.9—4.1	1.0—4.7	1.2—1.3	1.7	0.1—0.8	0.0—2.2	0.1—2.2	0.3—1.8	0.1—1.4	0.5—0.6	0.6—1.7
TiO_2	0.6—1.8	0.8—2.3	0.9—1.7	0.6—1.3	1.2	1.0—1.7	0.1—1.8	0.2—1.1	0.6—1.2	0.6—1.3	0.9—1.8	1.2—4.7
P_2O_5	0.6—0.46	0.04—0.61	0.13—1.8	0.8—1.3	0.2	0.01—2.8	0.0—1.3	0.0—1.7	0.02—0.76	0.1—1.4	0.21—0.51	0.18—2.6
SO_3	0.6—4.7	0.2—6.6	0.8—3.0	10.0—10.3	—	0.2—15.1	2.0—23.0	2.0—29.0	2.4—26.2	1.8—8.6	14.4—16.1	1.0—10.5
No. samples	13	28	12	2	10	12	125	93	7	10	2	6

Note: Type of coal: A, anthracite; B, bituminous; L, lignite; SB, subbituminous. Numbers in parenthesis refer to references. Values represent weight percentage of dry ash.

Table 3
AVAILABLE DATA ON RANGE IN CONCENTRATION OF TRACE CONSTITUENTS IN COAL ASHES

Element	Anthracites			High-volatile bituminous			Low-volatile bituminous			Medium-volatile bituminous			Lignites and subbituminous		
	Min	Max	Avg[a]	Min	Max	Avg[b]	Min	Max	Avg[c]	Min	Max	Avg[d]	Min	Max	Avg[e]
Barium	540	1,340	866	210	4,660	1,253	96	2,700	740	230	1,800	896	550	13,900	5,027
Beryllium	6	11	9	4	60	17	6	40	16	4	31	13	1	28	6
Boron	63	130	90	90	2,800	770	76	180	123	74	780	218	320	1,900	1,020
Chromium	210	395	304	74	315	193	120	490	221	36	230	169	11	140	54
Cobalt	10	165	81	12	305	64	26	440	172	10	290	105	11	310	45
Copper	96	540	405	30	770	293	76	850	379	130	560	313	58	3,020	655
Gallium	30	71	42	17	98	40	10	135	41	10	52	*	10	30	23
Germanium	20	20	*	20	285	*	20	20	*	20	20	*	20	100	*
Lanthanum	115	220	142	29	270	111	56	180	110	19	140	83	34	90	62
Lead	41	120	81	32	1,500	183	23	170	89	52	210	96	20	165	60
Manganese	58	365	270	31	700	170	40	780	280	125	4,400	1,432	310	1,030	688
Nickel	125	320	220	45	610	154	61	350	141	20	440	263	20	420	129
Scandium	50	82	61	7	78	32	15	155	50	7	110	56	2	58	18
Silver	1	1	*	1	3	*	1	1.4	*	1	1	*	1	50	*
Strontium	80	340	177	170	9,600	1,987	66	2,500	818	40	1,600	668	230	8,000	4,660
Tin	19	4,250	962	10	825	171	10	230	92	29	160	75	10	660	156
Vanadium	210	310	248	60	840	249	115	480	278	170	860	390	20	250	125
Ytterbium	5	12	8	3	15	10	4	23	10	4	13	9	2	10	4
Yttrium	70	120	106	29	285	102	37	460	152	37	340	151	21	120	51
Zinc	155	350	*	50	1,200	310	62	550	231	50	460	195	50	320	*
Zirconium	370	1,200	688	115	1,450	411	220	620	458	180	540	326	100	490	245

Note: An asterisk indicates insufficient data to compute an average value. Concentrations given in ppm.

[a] Number of samples used to compute average values equals 5, except cobalt (4).
[b] Number of samples used to compute average values equals 24, except tin (22) and zinc (14).
[c] Number of samples used to compute average values equals 8, except gallium (7) and tin (7).
[d] Number of samples used to compute average values equals 7, except cobalt (6), nickel (6), and zinc (6).
[e] Number of samples used to compute average values equals 13, except gallium (12), nickel (8), and scandium (10).

From EPRI Report CS-3122, Volume 1.

some of the organic matter. Fly ash, which accounts for the majority of ash generated, is the small size particles (in the microns to submicron size range) carried out of the furnace with the flue gas. Bottom ash and boiler slag are the larger size particles (up to millimeters in size) which drop to the bottom of the furnace during coal combustion.

The formation of power plant ash takes place in the furnace. There are two types of furnaces in widespread use at power plants: pulverized coal-fired furnaces, and cyclone furnaces.

A pulverized coal furnace burns fine coal, with particle size passing a No. 200 sieve. The pulverized coal is aerated and transferred to the burners, located on the furnace walls, where combustion takes place at a temperature ranging from 1100° to 1400°C. Pulverized coal furnaces can use either a "dry bottom" or "wet bottom" to collect and remove bottom ash from the furnace. A dry-bottom furnace produces 20 to 25% as bottom ash, which is collected in a dry hopper below the boiler. The remaining 75 to 80% is collected as fly ash. A wet-bottom furnace produces approximately 50% as bottom ash, which is collected in a water-filled hopper below the boiler. Approximately 50% of the total ash is collected as fly ash.

Table 4
PROXIMATE ANALYSES OF SOME NORTH AMERICAN COALS[10]

Rank and source	Moisture (percent) As mined	Moisture (percent) Dry basis*	Ash (percent) As mined	Ash (percent) Dry basis[a]	Volatile matter (percent) As mined	Volatile matter (percent) Dry basis[a]	Fixed carbon (percent) As mined	Fixed carbon (percent) Dry basis[a]	Heat value (Btu/lb) As mined	Heat value (Btu/lb) Dry basis
Lignite										
Montana	37.5	0	6.1	9.8	26.8	42.9	29.6	47.3	6,580	10,528
North Dakota	36.8	0	5.1	8.1	28.2	44.6	29.3	46.4	7,204	11,399
South Dakota	39.2	0	8.4	13.8	24.7	40.6	27.7	45.6	6,307	10,373
Texas	33.7	0	7.3	11.0	29.3	44.2	29.7	44.8	7,348	11,083
Ontario	46.7	0	7.4	13.9	30.2	56.7	16.0	30.0	5,280	9,906
Saskatchewan	33.5	0	6.7	10.1	24.0	36.1	35.8	53.8	7,576	11,392
Subbituminous										
Wyoming	23.4	0	3.6	4.7	33.6	43.9	39.4	51.4	9,382	12,248
Alberta	15.0	0	6.3	7.4	33.3	39.2	45.4	53.4	10,890	12,812
Bituminous										
Utah	7.5	0	5.6	6.1	39.7	42.9	47.2	51.0	12,520	13,535
Illinois	7.9	0	9.1	9.9	40.7	44.2	42.3	45.9	11,527	12,515
British Illinois	6.7	0	11.3	12.1	34.8	37.3	47.2	50.6	11,690	12,529
Alberta	7.4	0	7.1	7.7	32.1	34.7	53.4	57.7	11,630	12,559
Pennsylvania	3.4	0	5.3	5.5	35.9	37.2	55.4	57.3	11,734	14,217
Alabama	3.9	0	7.5	7.8	35.1	36.5	53.5	55.7	13,343	13,884
Nova Scotia	4.0	0	8.4	8.8	32.2	33.5	55.4	57.7	13,340	13,896
Anthracite										
Pennsylvania	3.2	0	11.5	11.9	9.3	9.6	76.0	78.5	13,043	13,474

[a] Calculated value.

The cyclone furnace burns coal crushed to a size less than 1.3 cm or 4 mesh, at a very high temperature. The coal is burned by continuous swirling in a high-heat-intensity zone, resulting in the fusing of ash into a glassy slag which falls to the bottom of the furnace. Between 70 to 85% of the ash is tapped from the furnace as boiler slag, leaving 15 to 30% of the ash to exit in the flue gas as fly ash.

The flue gases from both a pulverized-coal and a cyclone furnace exit the furnace and ultimately are discharged through a stack to the atmosphere. The flue gas is cleaned using one or several technologies to reduce the air-quality degradation. Flue gas cleaning (FGC) includes both removal of fly ash and, when needed, removal of sulfur oxides. The removal of sulfur oxides produces scrubber sludges.

C. Collection of Coal Ashes

There are four types of fly ash collection systems commonly used to capture ash particles:

1. Electrostatic precipitator (ESP)
2. Fabric filter (FF) or Baghouses
3. Wet scrubber
4. Mechanical collector

As seen in Table 5, the ESP is the most widely used fly ash collection system.[13]

Very few mechanical collectors are in use and, therefore, make up less than 1% of the total. Use of fabric filters is increasing, representing a growing application of a new technology in the electric utility industry, particularly with dry SO_2 control systems. The following is a brief discussion of the types of fly ash collection systems.

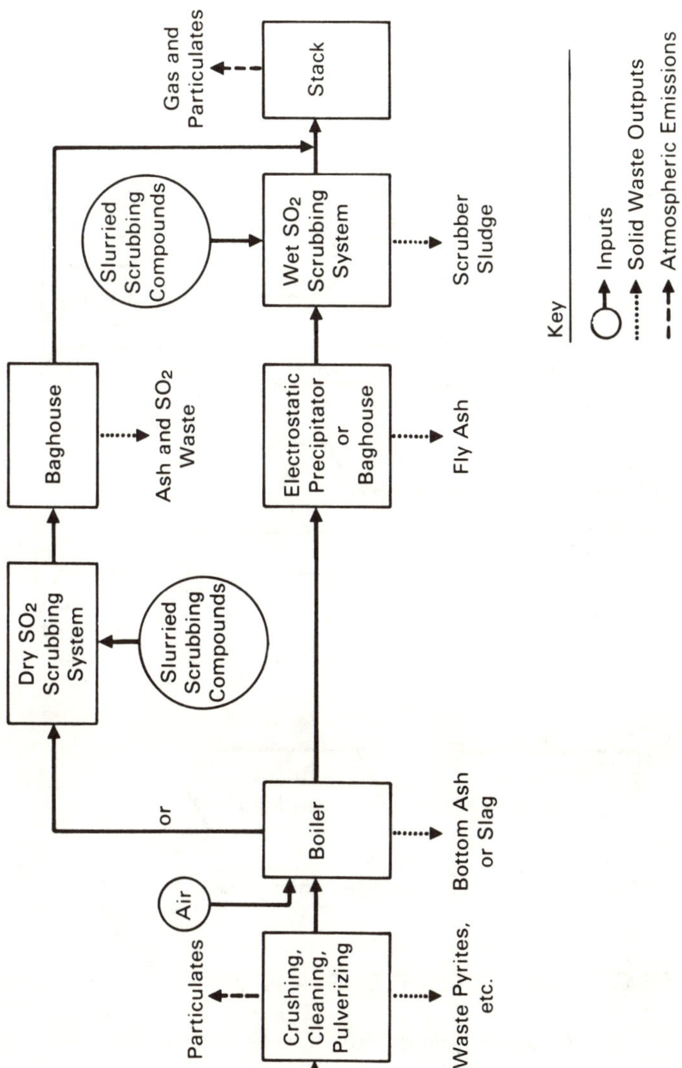

FIGURE 2. Schematic of high-volume solid waste production streams.

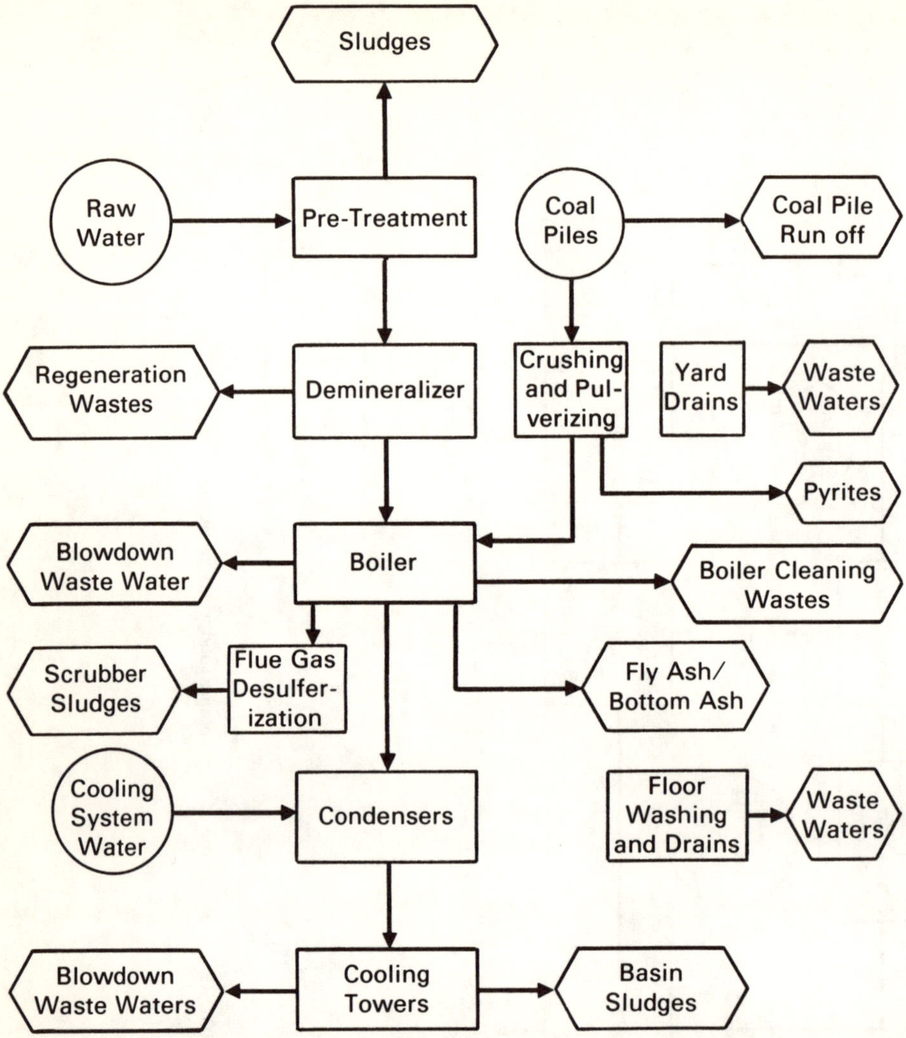

FIGURE 3. Schematic of low-volume waste production streams.

Table 5
FLY ASH COLLECTION SYSTEMS — 1979[13]

Type of system	Capacity of units (MW)	Fraction of total (%)
Electrostatic precipitator	196,717	95
Fabric filters	1,598	1
Wet scrubbers	8,685	4

1. Electrostatic Precipitators

As shown in Figure 4, an ESP typically is equipped with charging electrodes, wires or rigid frames, and collecting plates carrying DC charges of opposite polarity.[14] The electrodes impress a charge on the particulates in the flue gas flowing through the ESP, and the fly ash particles are attracted to the collecting plates. Periodically, the plates are vibrated or rapped, and the collected fly ash falls into hoppers located at the bottom of the ESP.

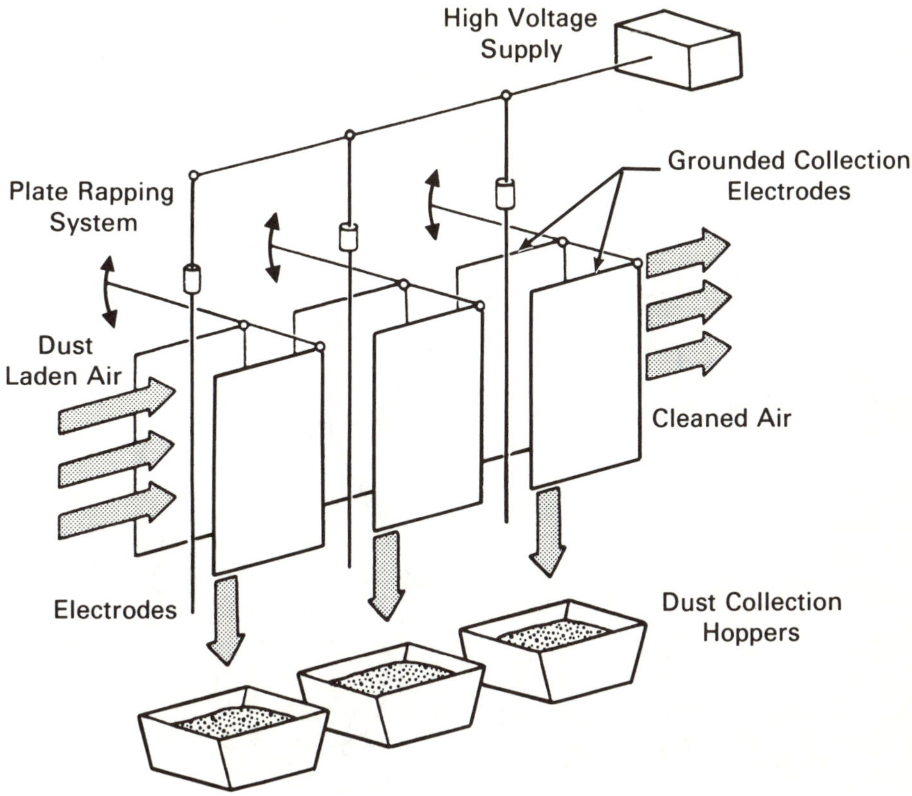

FIGURE 4. Schematic of an electrostatic precipitator (ESP).

ESPs have good overall efficiency, but do not collect particulates under 10 μm in size as well as fabric filters or wet scrubbers.[15] Collection efficiencies of 99 to 99.5% are common in newer installations. Difficulties have been experienced with ESP systems when low-sulfur coal is burned, because of the high resistivity of the ash. Some success has resulted from the installation of the ESP ahead of the air preheaters. These are known as "hot side" ESPs, as compared to "cold side" ESPs, which are installed after the air preheaters.

2. Fabric Filters

Fabric filters are being used by utilities in increasing numbers for several reasons:

1. They have been proven in full-scale installations.
2. The capital cost is lower than that of an ESP.
3. They are efficient in removing fine particulates.
4. Fabric filters are required when dry type flue gas desulfurization (FGD) systems are also used.

As shown in Figure 5, fabric filters are installed in enclosures commonly known as baghouses.[16] The gas path runs from the inside of the bag to the outside, with the fly ash being trapped by the fabric and the layer of collected ash. The efficiency of removal increases as the ash layer gets thicker. Periodically, the bags are cleaned by bypassing the gas flow around the compartments in turn and shaking the bags, applying a reverse air flow, or a combination of both. The displaced ash falls into collecting hoppers at the bottom of the bags.

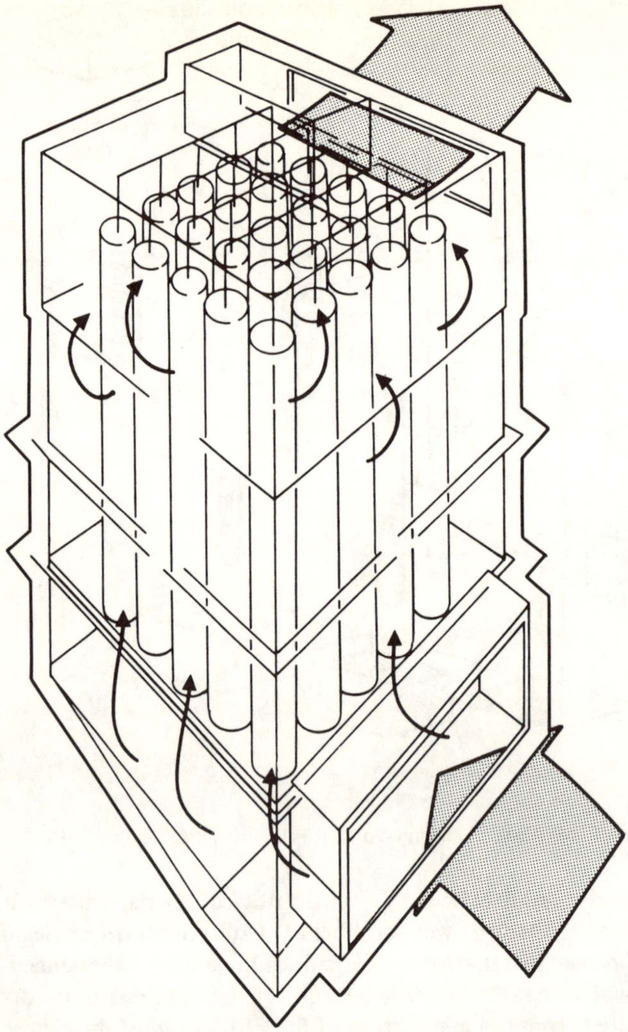

FIGURE 5. Schematic of a fabric filter (FF) or baghouse.

3. Wet Scrubbers

Wet scrubbers are no longer being installed primarily for fly ash removal from flue gas because:

1. Scrubber sludge is more difficult to handle than dry fly ash.
2. Some SO_2 is incidentally removed with the fly ash, resulting in low pH in the scrubber liquor which may then corrode metal components.
3. Mobile bed scrubbers become clogged with particulates.
4. Venturi scrubbers require a high pressure drop for efficient particulate removal.
5. Reuse of wet fly ash is more difficult.

Wet scrubbers are effective for final particulate removal when preceded by an ESP, as in the case of some FGD systems.

4. Mechanical Collectors

Mechanical collectors (cyclones) have generally fallen into disfavor because of low efficiency, especially in removing smaller particles. Previously, they were mostly used in combination with, and preceding, an ESP. However, even greater efficiencies are now obtained with either high-performance ESPs or FFs without mechanical collectors.

D. Collection of SO_2 Removal Wastes

There are three types of SO_2 removal and waste collection systems: wet, dry, and recovery.[17] The recovery system has had limited application in the U.S. and is not discussed here. Provided in Appendix B is a listing of the plants with SO_2 removal equipment and the type used.

1. Wet Systems

This type of FGD system is often referred to as a wet scrubber and is the most widely used SO_2 removal system in the U.S. The wet systems are distinguished by the reagent, or sorbent types used for SO_2 removal, and the fact that all produce a wet sludge. Various scrubber configurations are used to optimize contact between the sorbent and the SO_2 in the flue gas. The most popular and widely used type of scrubber is the spray tower. The wet systems also remove particulate material; however, fly ash adds to the quantity of sludge to be handled, and the trend is to remove fly ash in an ESP upstream of the SO_2 scrubber.

Wet scrubbers involve the continuous withdrawal of a bleed stream carrying the reaction products and varying amounts of fly ash. The systems use an alkali in water solution and/or suspension as a reagent for SO_2 removal. The alkali can be lime, limestone, alkaline fly ash, or in the case of double alkali scrubbers, a sodium salt. When alkaline fly ash acts as part or all of the reagent, it is either permitted to enter the scrubber with the gas or it is removed by an ESP, then slurried with water and reintroduced into the scrubber.

As shown in Figure 6, for the wet lime/limestone system, the flue gas enters the scrubber vessel where it contacts the recycling slurry which carries the reagent.[18] The slurry then drains to the bottom of the vessel, or to a reaction tank, where a bleed stream is separated and fresh reagent is added. The slurry then recycles to the scrubber vessel. The bleed stream, or blowdown, goes to a dewatering pond, or in most cases, to a thickener which is followed by a second step of dewatering prior to its reuse or disposal.

The FGD reaction products from wet systems are carried from the scrubber in a water slurry along with fly ash in varying amounts. The solids content varies from about 5 to 15% in the bleed stream. This can be increased to about 30 to 40% in a thickener and then to 50 to 80% by a vacuum filter or centrifuge for disposal or reuse. Chemical properties affect the extent to which a sludge can be dewatered.

2. Dry Systems

The growing interest in dry systems has resulted primarily from the ease of handling and lower costs. The dry system produces a practically dry powder (less than 1% moisture content), containing both fly ash and the sulfur compounds formed with the reagent. A sodium compound can be used as a reagent, although lime is the preferred reagent.

As shown in Figure 7, there is no upstream removal of fly ash in dry systems.[18] The sorbent solution contacts the flue gas in the reaction vessel or spray dryer; some of the fly ash and reaction products drop out at this point. Additional reaction occurs on the fabric filters, and most of the byproduct is removed. The dry product is recycled and/or removed from the system. Most current installations use bag houses, although an

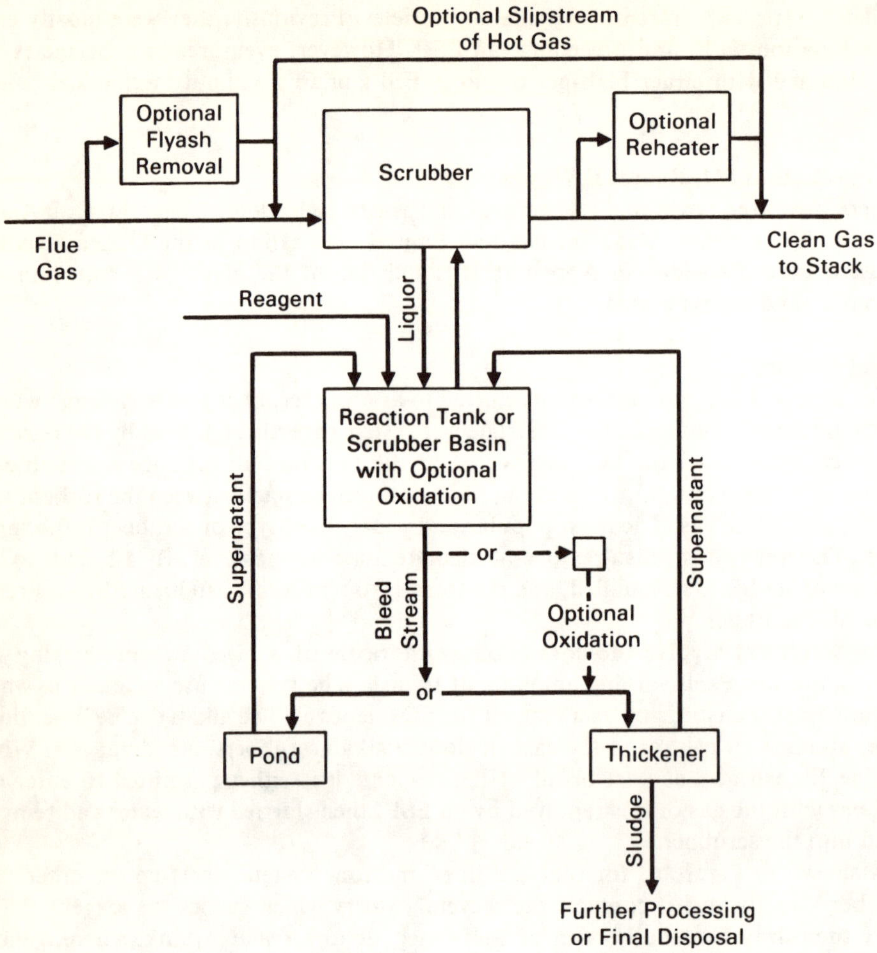

FIGURE 6. Schematic of a lime/limestone wet FGD system.[18]

ESP can also be used for removing the combined fly ash and reaction products from the gas stream.

E. Generation of Low-Volume Wastes

Low-volume wastes originate from a number of systems supporting the operation of a coal-fired power plant. Figure 3 provides an idea of how these wastes are produced in a power plant.[19] Each of these waste streams is site- or plant-specific except in a general sense. Many of these low-volume wastes are wastewater streams containing less than 10% solids by weight. The low-volume wastewaters are often codisposed with high-volume wastes, particularly when ponds are used for such disposal.

1. Boiler Blowdown

Boiler blowdown is the purge or bleed-off stream which is discharged from the water/steam cycle of a boiler. Steam from the boiler exhausts its thermal energy by passing through the turbine, where it reverts back to the liquid state in the condenser and is passed again to the boiler where the process is repeated. In the course of multiple recycling through the system, impurities (mostly salts) are concentrated due to water treatment additives, corrosion/erosion products from materials of construction for the turbine, boiler and ancillary piping, or small quantities of materials entering the steam

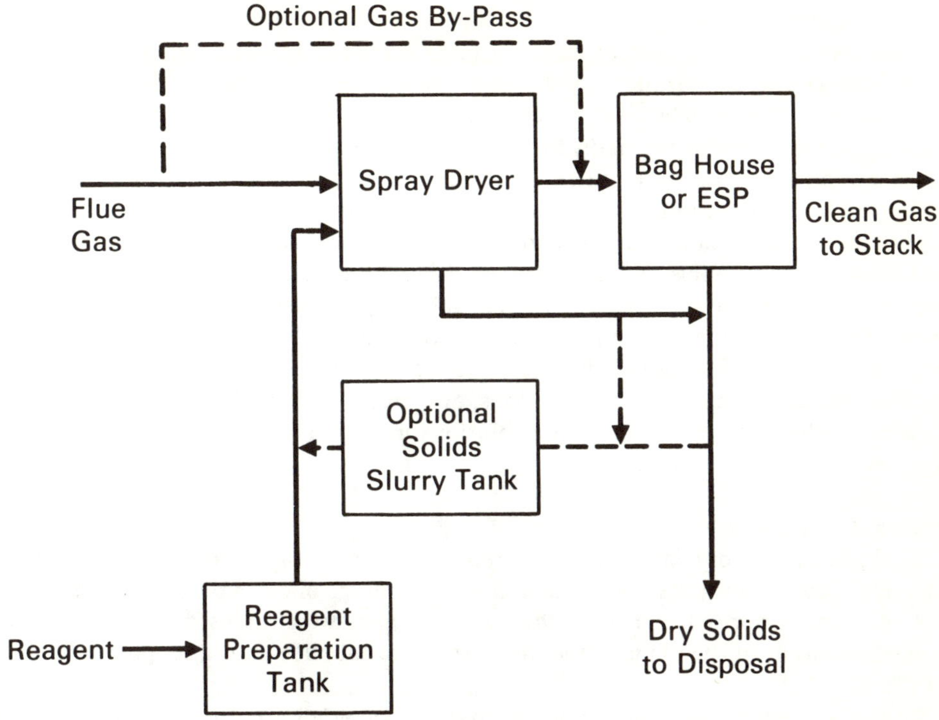

FIGURE 7. Schematic of a spray dryer FGD system.[18]

system from leaks in the condenser. The boiler is periodically blown down, or purged, such that waters containing relatively high concentrations of impurities can be discharged and replaced by high-quality water.

2. Demineralizer Regenerant

Demineralized water is required for the makeup to the steam cycle of a power plant. Typically demineralizers employ regenerable, polymeric resins for ion exchange. Varied resins are available for exchange of both anionic and cationic species.

A point in the ion exchange treatment process is reached where the resins must be regenerated with sulfuric acid for cationic exchangers and sodium hydroxide for anionic exchangers, including several rinse sequences with water. This regeneration sequence generates wastewaters.

3. Condenser Cooling System Blowdown

Condenser cooling system blowdown is the discharge, or bleed-off, wastewater from a power plant's main cooling system. This wastewater could originate from any closed-cycle cooling system such as towers, spray ponds, etc. The blowdown is required to maintain a suitable chemical operating condition with respect to scaling or corrosion in the cooling system.

4. Cooling Tower Basin Cleaning

In the operation of a cooling tower, the same water is passed through the condensers several times. Cooling tower basins provide temporary storage areas for this recycling water. Solids become concentrated in the cooling system water and gradually sludge builds up at the bottom of the basins. Periodically, these basins are cleaned mechanically to remove accumulated solids/sludges.

5. Metal Cleaning Wastes

There are two major types of metal cleaning wastewaters — fireside and waterside. Fireside cleanings are normally produced by water washing, without the use of chemical additives. The fireside wash waters contain some bottom ash or fly ash. Fireside washing is necessary to maintain the heat transfer and gas flow design levels of the furnace and downstream equipment, i.e., superheater, economizer, and air preheaters.

Waterside metal cleaning wastes is the term used to describe the various solutions containing chemical additives used to remove waterside buildup and scale in boilers and associated components. The chemical characteristics of the solutions used vary, depending on the equipment being cleaned and the specific cleaning needs of the equipment. Typical waterside chemical cleaning solutions include alkaline cleaning solutions, acid cleaning solutions, alkaline chelating or passivating rinses, organic agents, alkaline degreasers, ammoniated solutions, and proprietary compounds. For any given cleaning, there may be multiple rinse solutions used to assure removal of deposits or residues of cleaning chemicals.

6. Coal Pile Runoff

Coal pile runoff, or coal yard runoff, refers to surface runoff from the power plant's coal pile. Sometimes there is more than one coal storage pile at a plant. The active coal pile is often comprised of a five- to ten-day supply, which is used in supplying the immediate needs of the plant, while the long-term coal pile stores a supply for 90 days or more.

Coal pile runoff is generated when rainfall (or snowfall after melting) moves down through the coal and dissolves or washes into the suspension of certain materials present in or on the coal. The composition of the coal pile runoff depends on the characteristics of the coal being used at the site, the porosity of the coal pile, and climatological factors such as rainfall frequency and total amount.

7. Pyrites

Pyrites are best described as a mineral fraction of coal, i.e., rejects. They are separated at coal preparation facilities before shipment to the power plants, or at the power plant site, during the various stages of handling and pulverizing of the coal before introducing it to the furnace. Depending on coal source and coal preparation, a power plant could have substantial or very small amounts of pyrites to be disposed of.

8. Yard and Sump Drains

Power plants are served by a system of floor, yard, and equipment drains that collect a wide variety of effluents, including runoff due to incident rainfall, seepage from ground water sources, washdown wastewaters, bearing seals, and small equipment cooling waters and other waste flows. Depending on site-specific circumstances, sumps may collect oily and nonoily waters from various floor areas where small pieces of equipment are cleaned, from small equipment buildings and sheds, and from under large pieces of equipment such as boilers and turbine generators.

IV. QUANTITIES OF WASTES

In the U.S. there are over 400 power plants, generally comprised of multiple units which use coal as the principal fuel for producing electicity. These plants vary in size from under 20 MW(e) to about 3000 MW(e) in capacity. More than one unit is to be found at most of these power plants.[4] The smaller size plants were built in the 1940s and, progressively, the plant size increased as technological developments offered im-

Table 6
ANNUAL SOLID WASTE PRODUCTION AND COAL BURNED BY ELECTRIC UTILITIES, 1970 TO 1983[1-3]

Year	Coal consumption	Total solid wastes	Ash collected: Fly ash	Ash collected: Bottom ash	Slag	FGD sludge
1970	320.8	39.2	26.5	9.9	2.8	0
1971	327.9	42.8	27.8	10.1	5.0	0
1972	351.1	46.3	31.8	10.7	3.8	0
1973	388.2	49.3	34.6	10.7	4.0	0
1974	392.3	59.5	40.4	14.3	4.8	0
1975	406.0	60.0	42.3	13.1	4.6	0
1976	448.1	61.9	42.8	14.3	4.8	0
1977	477.2	67.8	48.5	14.1	5.2	0
1978	482.0	68.6	48.3	14.7	5.1	0.5
1979	518.5	76.2	57.5	12.5	5.2	1.0
1980	563.5	68.3	48.3	14.5	3.6	1.9
1981	596.8	71.4	50.3	12.9	5.2	3.0
1982	593.7	69.4	47.9	13.1	4.4	4.0
1983	625.2	75.0	52.4	14.0	3.9	4.7

Note: Values given in million short tons.

proved economies of scale. Figure 1 shows the distribution of the power plants,[3] and Appendix A lists them by state, starting dates, number of units, coal type burned, and capacity for electricity generation.[4]

A. High-Volume Wastes

The quantities of the solid wastes produced vary according to the types of coal used, the configuration of the boilers, and the types of flue gas desulfurization (FGD) and particulate control systems used.

Coal-fired power plants produce large quantities of fly ash, bottom ash, boiler slag, and FGD sludge. FGD sludge is a more recent solid-waste product because of increased use of FGD systems on new units to meet the New Source Performance Standards beginning September 1978. Some older units are being retrofitted with FGD systems. Table 6 shows the annual tonnage of coal burned[20-21] and the solid wastes produced nationally in the electric utility industry.[22] Table 7 gives the aggregate quantities of fly ash, bottom ash, and FGD sludge produced by state in 1981.[23]

In 1983, in the aggregate, the coal-fired power plants produced a total of about 75 million tons of fly ash, bottom ash/slag and FGD sludge. Of this amount, 52.4 million tons were fly ash, 14.0 million tons were bottom ash, 3.9 million tons were boiler slag, and 4.7 million tons were FGD sludge. These generation tonnages are expected to increase to about 180 million tons annually by the year 2000. These projections are based on correlations from historical data on coal use in the electric utility industry and the amount of fly ash, bottom ash, and slag produced.

As seen in Table 6, during the 1970s, a steady average annual increase of about 7.6% in ash collection was observed. Part of the increase in ash generation is reflected in the increased efficiency and reliability of flue gas cleaning equipment upgraded or placed into service during the 1970s. Variations in coal ash content from year to year also can contribute to fluctuations in ash generation. In the 1980s, the increase in coal burned by the utilities has fallen to 2.7% per year.

Table 8 shows the summary of data collected by the authors from a sample of 68

Table 7
COAL CONSUMPTION, TOTAL CAPACITY, AND SOLID WASTE QUANTITIES PRODUCED IN 1981, BY STATE

State name	Number of plants	Total capacity MW(e)	Amount of coal burned[a]	Total amount waste produced[a]: Fly ash	Bottom ash/slag	FGD sludge
Alabama	10	10,263	19,452	2,064	446	0
Arkansas	2	2,158	5,958	264	84	0
Arizona	4	4,726	14,004	1,326	344	21.4
Colorado	10	4,307	11,353	651	190	10.2
Delaware	2	782	1,837	215	35	0
Florida	8	4,429	9,210	989	282	0
Georgia	10	10,350	21,995	1,913	495	0
Illinois	28	16,672	32,442	1,632	1,338	323
Indiana	25	15,369	29,957	2,154	981	50
Iowa	21	4,998	11,073	638	184	0
Kentucky	21	13,266	24,364	1,879	1,057	160
Kansas	7	4,449	11,260	453	496	392
Louisiana	3	1,130	1,225	43	10	0
Maryland	5	3,142	5,196	646	204	0
Massachusetts	3	1,590	836	158	14	0
Michigan	26	11,080	22,755	1,691	335	0
Minnesota	21	4,655	11,834	834	257	54
Mississippi	5	2,380	3,329	270	57	26
Missouri	20	11,875	23,314	1,414	1,146	29
Montana	3	889	3,119	345	64	0
Nebraska	7	3,060	5,058	237	74	0
Nevada	3	2,155	4,894	495	215	26
New Hampshire	1	459	886	56	9	0
New Jersey	6	1,693	2,651	128	99	0
New Mexico	3	3,881	10,596	1,835	449	11
New York	10	2,371	5,652	739	173	0
North Carolina	13	11,762	20,183	2,307	585	0
North Dakota	8	3,241	12,600	566	420	128
Ohio	33	22,330	48,221	5,117	1,531	680
Oklahoma	4	3,040	7,836	309	70	0
Oregon	1	561	1,197	43	21	0
Pennsylvania	26	18,922	38,755	4,724	1,019	533
South Carolina	9	4,125	8,669	760	165	12
South Dakota	3	456	2,319	56	128	0
Tennessee	8	10,090	20,491	1,859	575	0
Texas	12	13,556	52,701	3,872	1,596	521
Utah	7	2,157	4,887	419	107	0
Vermont	1	—	14	1	1	0
Virginia	7	3,644	6,543	648	170	0
Washington	1	1,330	4,964	468	312	0
West Virginia	12	14,992	30,046	3,326	953	0
Wisconsin	22	6,338	11,737	851	268	0
Wyoming	7	5,240	13,728	931	543	61
Total U.S.	438	263,918	579,147	49,523	17,556	3,038

Note: Data derived from FERC and EEI.

[a] Values given are in thousands of tons.

Table 8
TRENDS IN FLY ASH AND BOTTOM ASH PRODUCTION BASED ON A SAMPLE OF PLANTS

Year	Sample size: Total capacity (MW)	Sample size: # of Plants	Average production per plant[a]: Fly ash (FA)	Average production per plant[a]: Bottom ash (BA)	Ratio FA/BA	Average production per megawatt capacity[a]: Fly ash (FA)	Average production per megawatt capacity[a]: Bottom ash (BA)	Ratio FA/BA
1970	15636	21	140.5	61.5	2.29	0.192	0.086	2.23
1971	16302	14	264.8	139.5	1.90	0.227	0.120	1.89
1975	59936	54	224.1	93.8	2.39	0.202	0.088	2.29
1977	63737	57	234.4	99.1	2.36	0.210	0.092	2.28
1979	71773	68	222.5	90.2	2.47	0.211	0.086	2.45
1981	64159	62	241.2	94.9	2.54	0.233	0.095	2.45
1983	58116	58	207.1	80.3	2.58	0.207	0.079	2.62

[a] Values given in thousands of short tons.

power plants in the U.S. These data show that the average per-megawatt production of fly ash has remained relatively constant during the 1970s even though the average per plant fly ash production showed a small decline. The bottom ash production per megawatt and per plant declined slightly during this same period. This trend in production is further substantiated by the ratio of fly ash to bottom ash production (Table 8). This ratio has gradually increased between 1970 and 1983. Larger fly ash production may be due to increased collection efficiencies in ESPs and FFs. Data from several individual power plants show a declining trend in the production of fly ash and bottom ash with time. Only a few plants in this sample show an increase in fly ash and bottom ash production with time.

For the 1970s, the ash content of coal burned by utilities had a weighted mean value of about 13.8%, based on U.S. Department of Energy (DOE) and National Ash Association (NAA) data.[23] This 13.8% is compared to the 1980s weighted mean of 11.56%. This means that today every ton of coal burned is generating about 17% less ash than in the 1970s. In fact, coal receipts for 1983 had an average weighted mean ash content of 10.4%.[24]

The explanation for this decrease is most likely twofold. First, certain utilities are shifting coal sources to obtain low sulfur or low ash content coal. This shift is being done to bring the facility into compliance with applicable air quality regulations with regard to sulfur oxide emissions and particulate emissions, without requiring major investment in flue gas cleaning equipment. Secondly, coal is more often washed to remove dirt, adsorbed sulfides, and other noncombustible materials at mine sites before it is shipped to power plants.

B. Low-Volume Wastes

The following is a discussion of the quantities of wastes which can be generated from low-volume sources.

1. Boiler Blowdown

The major waste stream from the boiler system is the boiler blowdown. The blowdown rate is generally in the range of 0.1 to 3% of the steam flow, and it is either continuous or intermittent. The boiler blowdown wastewaters contain suspended and dissolved solids as well as scale constituents formed in boiler water, boiler tube corrosion products, and internal chemicals added to boiler water.

Table 9
TYPICAL WATER TREATMENT WASTEWATER FLOWS

Process	Range of flows (gal/1000 lb water treated)
Clarifier blowdown	1—4
Lime-soda	1—4
Raw water filtration backwash	0—6
Feed water filter	0—6
Sodium zeolite regeneration	0.5—3
Cation exchange regeneration	0.5—3
Anion exchange regeneration	0.5—3
Evaporator blowdown	12—40
Condensate filtration and ion exchange	0.02—0.6
Condensate powder	0.01—0.06

From U.S. EPA, Development Document for Effluent Limitation Guidelines and New Source Performance Standards for Steam Electric Power Generating Point Source Category, EPA 440/1-74/029-a, 1974.

2. Water Treatment

Raw water has to be treated prior to its use as make-up water in the boiler. The treatment operations can be divided into (1) pretreatment step for removing hardness, silica, and suspended solids in the preparation of higher-quality water, (2) filtration — or special adaptation for the removal of suspended matter, and (3) ion exchange and other techniques for the alteration or removal of dissolved solids by methods not involving chemical precipitation. The control of silica and sodium compounds is necessary to prevent their volatilization and/or carrydown from the boiler and subsequent deposition on turbine blades. Table 9 gives typical flow data.

3. Cooling Tower Blowdown

The blowdown flow depends upon the maximum allowable concentration for each of the dissolved and suspended species in the circulating water without risking undue scale formation and corrosion in the cooling systems. The blowdown rate typically ranges between 0.5 to 3% of the recirculating water flow.

4. Cooling Tower Basin Cleaning

Depending upon the quality of the make-up water used in the cooling tower, carbonates can be deposited in the tower basin. Silt and sand also accumulate there. Similarly, depending upon the inefficiency of chlorine dosages, some algae growth may occur on basin walls. Some debris carried in the atmosphere may also collect in the basin. Consequently, sludge is removed periodically with a front-end loader and truck.

5. Metal Cleaning

The quantity and quality of metal cleaning wastes are determined by the plant operating and maintenance procedures, as well as by the cleaning practices followed by the utility.

6. Boiler Tube Cleaning

The value of wastes resulting from boiler tube cleaning operations can vary between 3 to 10 times the boiler volume of about 30,000 to 100,000 gallons, depending upon the specifics of the application. The boiler fireside is cleaned with high-pressure alkaline water containing sodium salts such as soda ash, caustic soda, and/or phosphates. The frequency of cleaning boiler tubes and boiler fireside varies from twice a year to once every eight years. The typical cleaning frequency is once every three years.

7. Air Preheater Cleaning

Soot and fly ash accumulate on the preheater surfaces and the deposits must be removed periodically to maintain good heat transfer rates as well as to avoid plugging of the tubes by metallic elements. Preheaters are cleaned with high-pressure fire hoses. The washing fluid may contain soda ash and phosphates or detergents which are added to neutralize excess acidity.

8. Miscellaneous Small Equipment Cleaning

Occasionally, other plant components such as condensate coolers, hydrogen coolers, air compressor coolers, stator oil coolers, etc., are cleaned chemically. Inhibited hydrochloric acid is the common chemical used for cleaning. Detergents and wetting agents are also added when necessary. The waste volume is, of course, smaller than that encountered in other types of chemical cleaning.

V. WASTE CHARACTERISTICS

In a recent EPRI report,[12] published data on physical-chemical composition of utility solid wastes were compiled. This report has been used in preparing the following description on physical-chemical composition of fly ash, bottom ash, and FGD sludge. Additional data, developed in EPRI research activities, have been used to augment this chapter.[25-30] Figure 8 shows the physical appearances of bottom ash, FGD scrubber sludge, and fly ash.

Limited data on physical properties, e.g., specific gravity, dry bulk density, specific surface area, permeability, and grain-size distribution, are available in the literature. The chemical composition data most commonly available are on the elemental composition of the bulk material. Nineteen inorganic elements (namely, Ag, As, B, Ba, Cd, Co, Cr, Cu, Hg, F, Mn, Mo, Ni, Pb, Se, Sr, Ti, V, and Zn) are most frequently measured in the chemical analyses of the wastes.[12]

In addition to the bulk chemical analysis data, laboratory leaching extractions have yielded a limited amount of data on the solubility of chemicals contained in the wastes. These laboratory data provide a basis for assessments on the potential for environmental releases of waste constituents when disposed of or reused.

A. Physical Properties of High-Volume Wastes

Several physical characteristics have been measured for selected samples of fly ash, bottom ash, and FGD sludges: most commonly, particle size distribution, specific gravity, dry bulk density, specific surface area, shear strength, and coefficient of permeability.

1. Particle or Grain Size Distribution

The high-volume solid wastes do not consist of homogeneous material of the same size particles. In fact, the particles vary in diameter from about 2 μm to over 20,000 μm. Bottom ash/slag consists of larger size particles whereas fly ash generally consists of smaller size particles. Figure 9 presents the ranges of particle size distributions for some fly ashes and bottom ash.[31] From the figure it is clear that there is some overlap in the particle sizes between fly ash and bottom ash. This overlap is in the particle size range of 70 μm to 600 μm. The greatest proportion of bottom ash particles lie in the

FIGURE 8. Photographs of coal ashes and sludges: (A) bottom ash, (B) dried FGD sludge, (C) fly ash, and (D) fly ash.

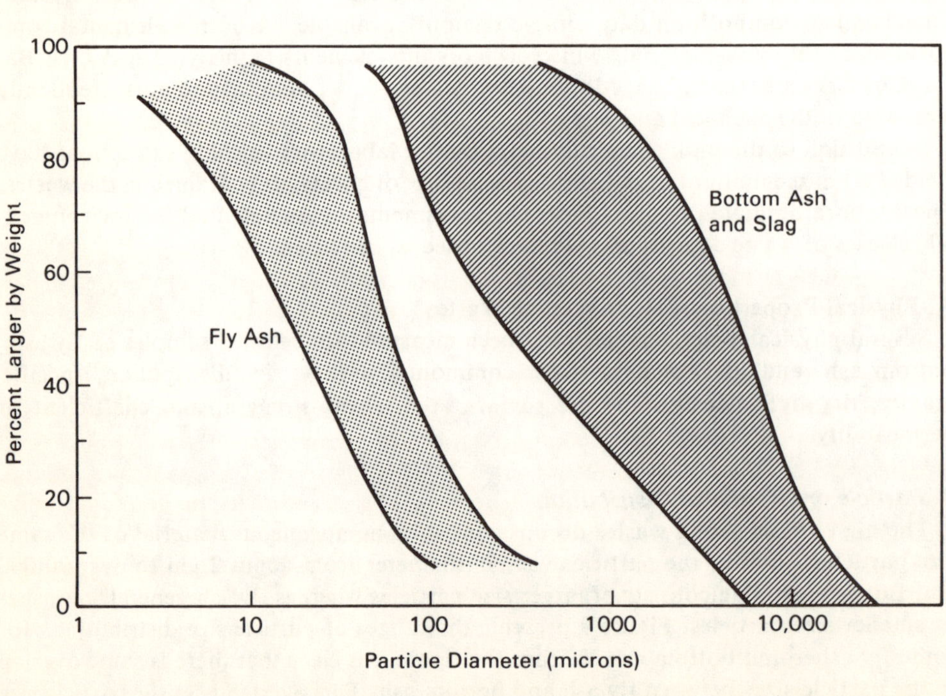

FIGURE 9. Particle-size distribution for fly ash and bottom ash.[12]

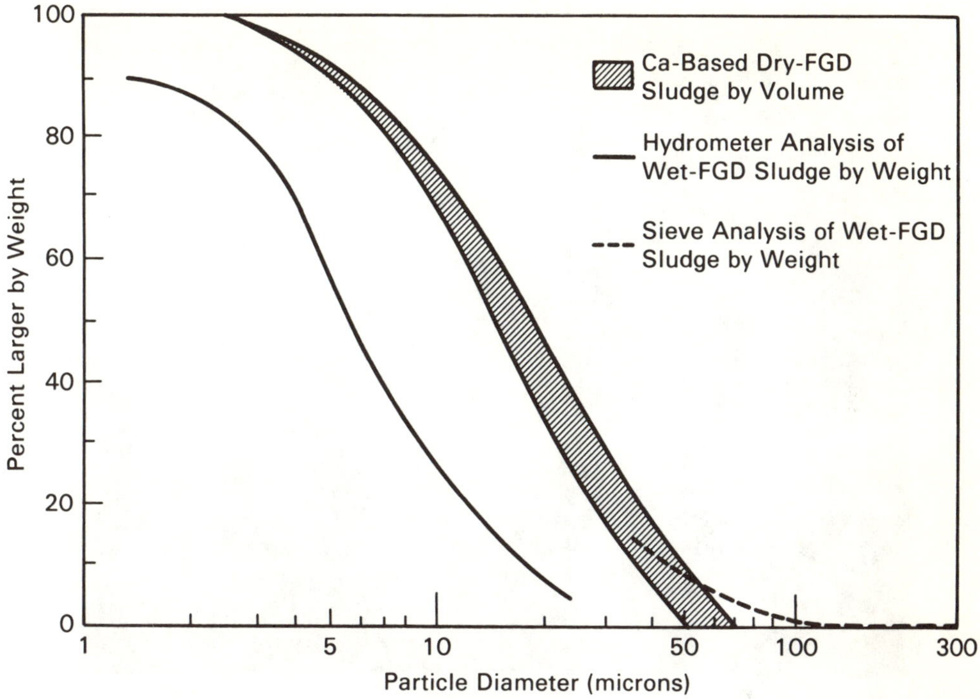

FIGURE 10. Particle-size distribution for FGD sludges.[12]

size range of 100 to 10,000 μm, whereas most of the fly ash particles are between 5 and 100 μm in size. In general, except for boiler slag, the bottom ashes are coarser than the fly ashes, and do not have the spherical shape of fly ash.

The very little data available on the particle size distribution of the FGD sludges indicate that the particles in a wet FGD sludge may be distributed in the range of 1 to 50 μm (Figure 10).[32,33] It appears that FGD sludges can vary markedly, because the wastes contain calcium-based reaction products, excess scrubbing reagents, and fly ash in proportions from a trace to more than 50% by weight.

2. Particle Morphology

Morphology of the waste particles has been examined by the use of scanning electron microscopy (SEM) and light optical microscopy (OM) techniques. Fly ash particles have been broadly categorized as (1) cenospheres, (2) plerospheres, (3) spherical particles with crystals, and (4) irregularly shaped solid particles.[26] SEM pictures are shown in Figure 11 to illustrate these morphological characteristics. The cenospheres are hollow, thin-walled spheres which can float on water. Smaller spheres can be incorporated in a larger size sphere producing what are known as plerospheres. Often the spherical particles also have platy or needlelike crystals adhering to their surfaces forming agglomerates. In addition, fly ash includes irregular-sized particles made of partially combusted carbon, magnetic particles, or glassy materials.

Morphological classifications for bottom ash, slag, and FGD sludges are not readily available.[18] Figure 12 shows the SEM pictures for the bottom ash. Figure 13 shows the SEM pictures for the FGD sludge.

3. Specific Gravity and Specific Surface Area

Reported specific gravities for the coal fly ash range between 1.6 to 3.1, and for bottom ash and slag range between 2.2 to 2.8.[12] Because FGD sludges are produced

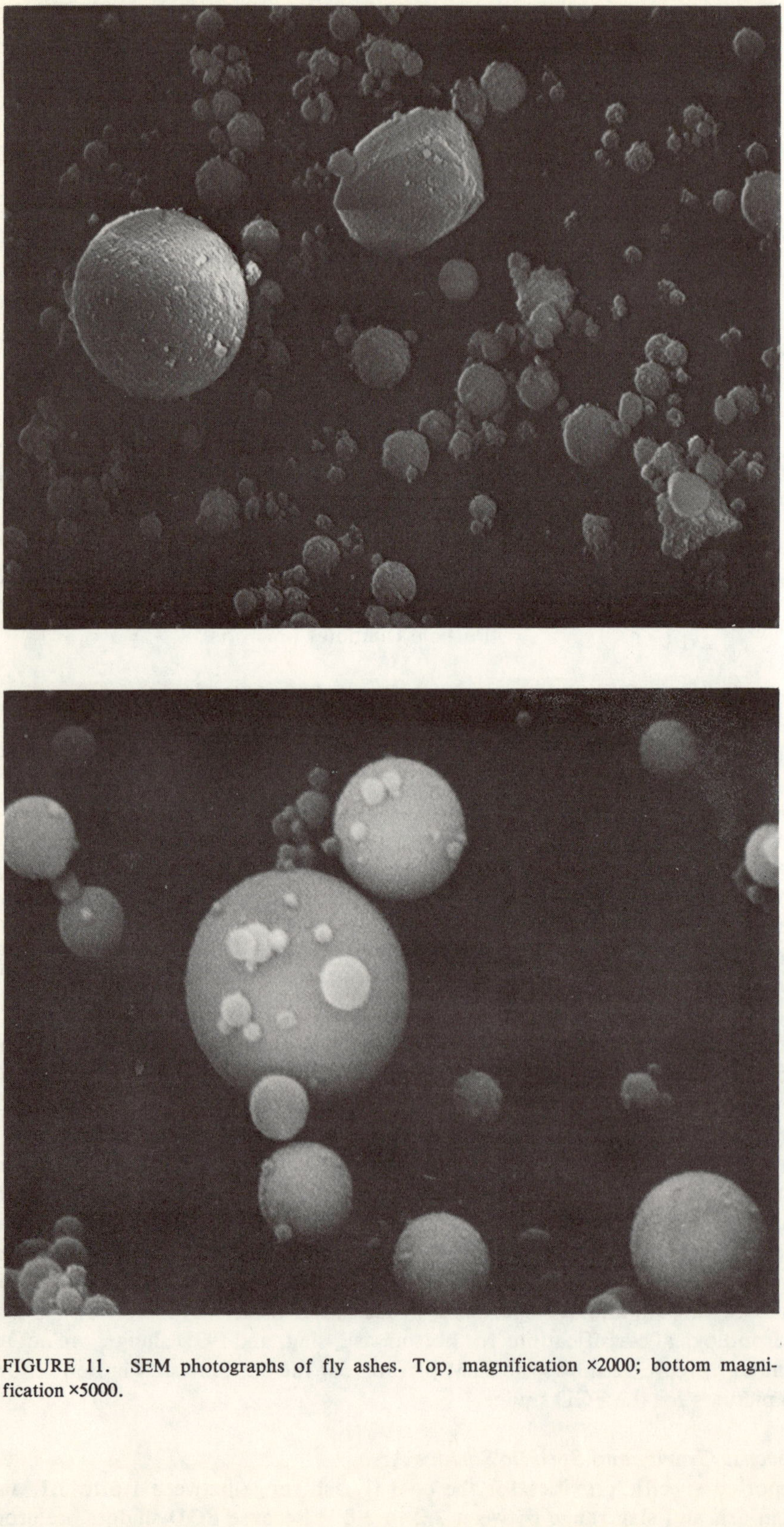

FIGURE 11. SEM photographs of fly ashes. Top, magnification ×2000; bottom magnification ×5000.

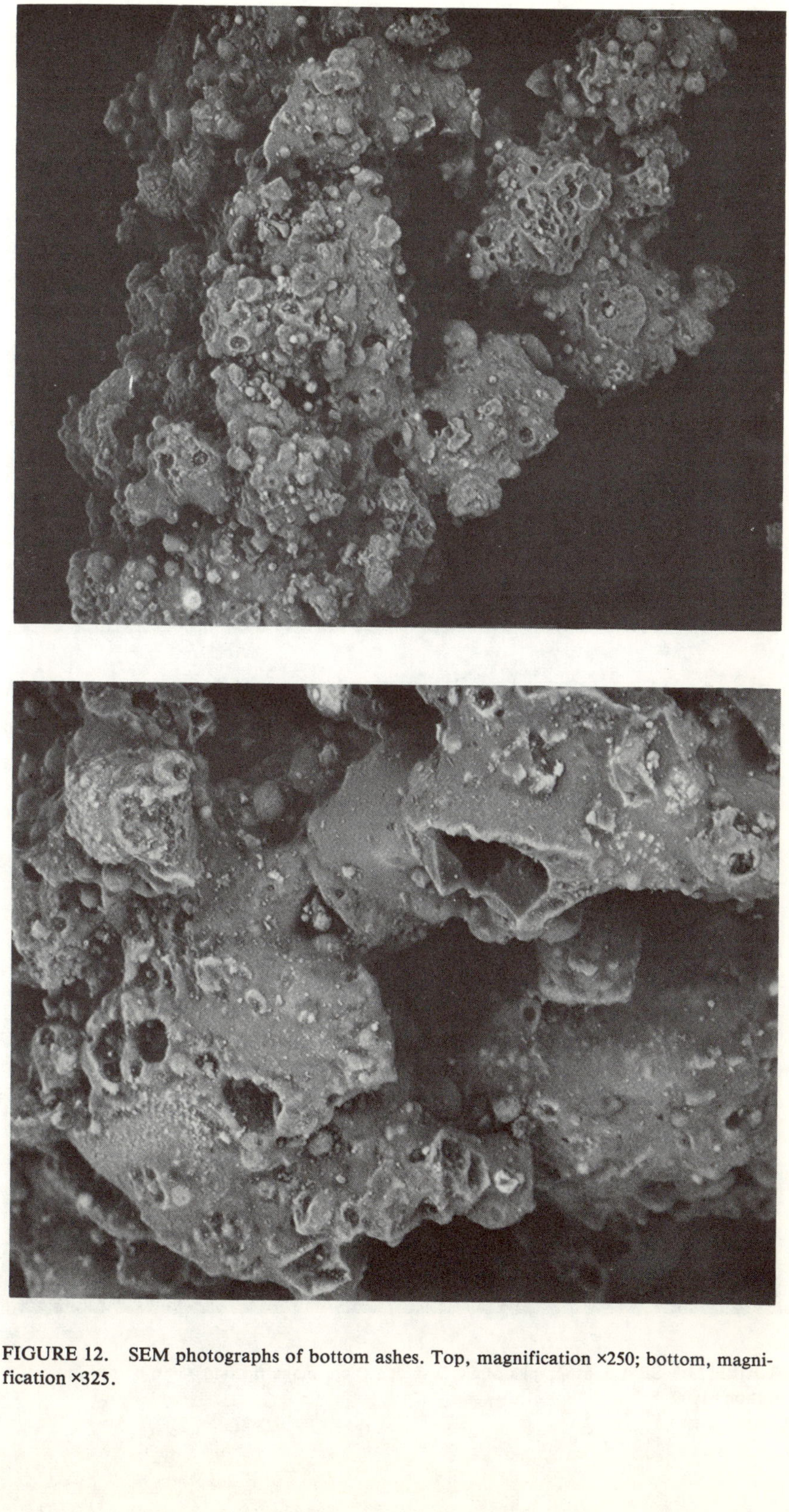

FIGURE 12. SEM photographs of bottom ashes. Top, magnification ×250; bottom, magnification ×325.

FIGURE 13. SEM photographs of FGD sludge. Top, magnification ×500; bottom, magnification ×1000.

from a variety of scrubbing systems, the specific gravities of the untreated wastes vary accordingly. A specific gravity value of 2.5 for the untreated FGD sludge solids has been most commonly reported.[34]

Specific surface areas for fly ash range between 0.2 to 0.8 m^2/g.[12] Bottom ash has specific areas of about 0.4 m^2/g.[35] Generally, the specific area is higher when ash is composed of smaller particles. In a recently completed study,[28] measured surface area for fly ash was 2.9 m^2/g which is much higher than the range reported earlier. A dry FGD process waste showed a surface area of 9.7 m^2/g.[28]

4. Dry Bulk Density

Dry bulk density is a physical property of solids which determines the strength, stability, and settling behaviors of the waste material. Reported dry bulk density of bottom ash and slag solids ranges between 46 to 99 lb/ft^3.[36,37] The comparable dry bulk density of coal fly ash solids ranges between 63 to 89 lb/ft^3.[36,38,39]

The bulk density of FGD sludge varies much more than fly ash and bottom ash due to large variations in the sludge collection and handling systems employed. Oftentimes the FGD sludges are chemically stabilized. The overall dry bulk density of wet FGD sludges is lower and varies in the range of 22 to 33 lb/ft^3. The dry SO_2 spray dryer FGD waste shows a range of 61 to 91 lb/ft^3 for dry bulk density.

5. Permeability

Another physical property commonly measured is permeability, an important physical characteristic which influences the movement of water through the waste materials. Most bottom ash material permeability lies in the range of 10^{-3} to 10^{-1} cm/sec. Fly ash material permeability is in the range of 10^{-7} to 10^{-4} cm/sec. It appears that fly ash permeability from lignite coals is around 10^{-7} cm/sec whereas permeability of fly ashes from bituminous and subbituminous coals are variable in the range of 10^{-7} to 10^{-4} cm/sec.[12,31]

The FGD sludge permeabilities are different depending on the treatment it may have received. The untreated FGD sludges show permeabilities of 10^{-5} to 10^{-4} cm/sec. However, various stabilization/fixation treatments tend to decrease the permeabilities to values between 10^{-7} to 10^{-6} cm/sec.

B. Chemical Characteristics of Coal Fly Ash, and Bottom Ash/Slag

A wide variety of chemical analyses of utility wastes have been made and reported in the literature. Typically a bulk sample of the waste material is obtained from a plant and total elemental analyses are carried out by Atomic Absorption (AA), Inductively Coupled Argon Plasma (ICAP), and Neutron Activation Analysis (NAA) methods.

1. Bulk Composition

Major inorganic compounds of Si, Al, Fe, Ca, and Mg account for over 90% of the chemicals in coal ashes because fly ash, bottom ash, and boiler slag are chiefly the unburned mineral matter that was present in the fuel coal. The elemental composition of the solid wastes is affected by whether the coal burned is bituminous, subbituminous, or lignite; whether the coal source is eastern, midwestern, or western; and whether the boiler type is cyclone or pulverized coal-fired. In addition, the chemical composition of the solids may change depending on the collection system used, namely electrostatic precipitator, mechanical collector, fabric filter, or wet scrubber.

Figure 14 gives a general composition as oxides of major elements of coal ashes derived from bituminous, subbituminous, and lignite coals.[40] All three types of coal ashes contain silicon dioxide as the major constituent along with approximately the

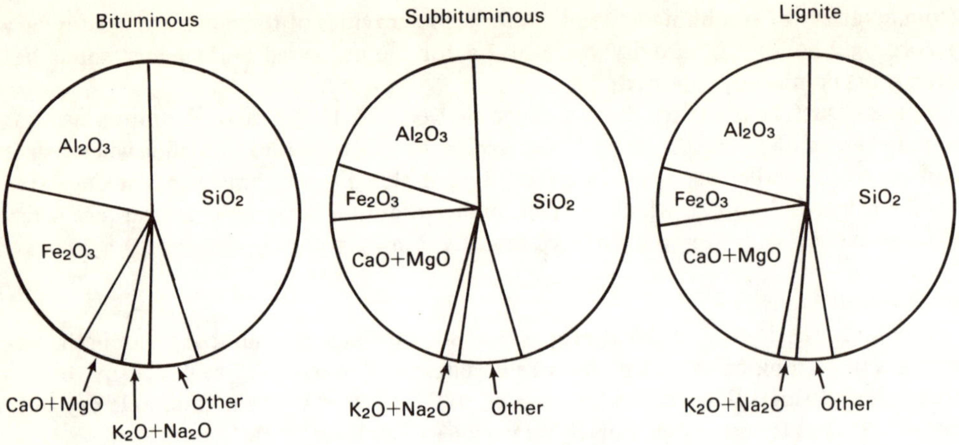

FIGURE 14. Major compounds constituting the coal ashes from three different coal types.[12]

Table 10
ELEMENT CONCENTRATIONS IN ASH FROM THREE GEOGRAPHIC SOURCES[12]

	Eastern coal			Midwestern coal			Western coal		
Element	n	Range	Median	n	Range	Median	n	Range	Median
Ag	14	0.25—80	0.695	12	0.10—1.10	0.39	20	0.040—6	0.26
As	39	2.0—279	75	40	0.50—179	54	48	1.3—129	18
B	18	10.0—580	121	19	10—1300	870	34	41.9—1040	311
Ba	52	52—2200	892	43	300—4300	905	34	300—5789	2700
Cd	41	0.10—8.24	1.59	25	0.50—18	2.6	43	0.10—14.3	1.01
Co	41	6.22—79	40.6	42	19—70	35.7	29	4.9—69	13
Cr	54	34—437	165	46	70—395	172	46	3.4—265	45
Cu	42	3.7—349	136	32	20—330	125	37	29—340	74.8
F	17	0.40—89	8.8	7	3.2—300	75	26	0.40—320	50.1
Hg	29	0.02—4.2	0.192	28	0.005—0.30	0.044	39	0.005—2.5	0.067
Mn	51	79—430	190	35	194—700	410	38	56.7—769	194
Mo	30	0.84—51	15	17	7.0—70	43	41	1.4—100	12
Ni	27	6.6—258	78	28	26—253	121	38	1.8—229	38
Pb	41	1.3—222	18	31	3.0—252	149	50	0.40—250	26.1
Se	48	0.36—19	8.05	32	0.08—19	7	42	0.13—19	4.1
Sr	36	59—2901	801	26	30—2240	423	15	931—3855	2300
Tl	11	7.0—28	25	24	2.0—42	16	7	0.10—3.50	1.06
V	37	110—551	269	31	100—570	270	35	11.9—340	94
Zn	45	16—1420	163	27	20—2300	600	43	4.0—854	71

Note: Ashes are of all types, from both types of furnaces. Values given in mg/kg.

same proportional amount of aluminum oxide. Iron oxide is relatively high in ashes from bituminous coal, and the oxides of alkali elements are present in smaller quantities in ashes from buituminous coal. Many of the other inorganic elements are found in the parts-per-million concentration range.

Table 10 shows the concentration ranges and the median values for Ag, As, B, Ba, Cd, Co, Cr, Cu, F, Hg, Mn, Mo, Ni, Pb, Se, Sr, Tl, V, and Zn reported for the ashes by coal source. These aggregate data have been broken down by the bottom ash, mechanical hopper ash, and fly ash in Table 11.

Table 11
ELEMENT CONCENTRATIONS IN THREE TYPES OF ASH

	Bottom ash			Mechanical hopper ash			Fly ash			
Element	n	Range	Median	n	Range	Median	n	Range	Median	Soil[a]
Ag	7	0.1—0.51	0.20	9	0.08—4	0.70	32	0.04—8	0.501	0.1
As	28	0.50—168	4.45	10	3.3—160	25.2	48	2.3—279	56.7	6
B	14	41.9—513	161	3	205—714	258	58	10.0—1300	371	10
Ba	27	300—5789	1600	14	52—1152	872	96	110—5400	991	500
Cd	23	0.1—4.7	0.86	13	0.40—14.3	4.27	77	0.10—18	1.60	0.2
Co	17	7.1—60.4	24	10	6.22—76.9	48.3	87	4.90—79	35.9	8
Cr	27	3.4—350	120	15	83.3—305	172	112	3.6—437	136	100
Cu	22	3.7—250	68.1	14	42.0—326	130	80	33.0—349	116	20
F	9	2.5—104	50	4	2.50—83.3	41.8	41	0.40—320	29	200
Hg	22	0.005—4.2	0.023	10	0.008—3.00	0.073	70	0.005—2.50	0.10	0.08
Mn	25	56.7—769	297	12	123—430	191	96	24.5—750	250	850
Mo	12	0.84—21	3.9	8	1.05—42.9	1.18	69	1.15—100	18	2
Ni	17	14—258	38	9	28.6—216	105	71	1.80—253	54	40
Pb	26	0.4—90.6	7.1	14	5.2—101	13	87	3.10—252	66.5	10
Se	25	0.08—14	0.601	16	0.13—118	5.52	86	0.60—19	9.97	0.2
Sr	5	170—1800	800	7	396—2430	931	73	30.0—3855	775	300
Tl	1	2.0	2.0	0	—	—	41	0.10—42	16.0	—
V	16	12.0—377	141	8	100—377	251	88	11.9—570	248	100
Zn	18	4.0—798	99.6	15	56.7—215	155	87	14.0—2300	210	50

Note: Values given in mg/kg. Data from all coal sources, both furnace types.

[a] Average soil composition from M.J. Dudas, personal communication, 1983.

These tables show considerable variability in the concentrations of the many inorganic elements measured in coal ashes. There are some observable differences in the chemical composition of the fly ash and bottom ash. Fly ash shows that most of the elements (with the exception of F, Sr, Ba, and Mn) are present in higher concentrations compared to bottom ash. Particular concentration trends were observable for Se and Pb, whose concentrations increased from bottom ash to mechanical collector ash to fly ash. This kind of concentration association has been hypothesized to be a consequence of the volatilization-condensation phenomenon.[12] Furthermore, the volatilized trace metals are deposited on small particles because such particles have larger surface areas. Consistent with this hypothesis As, Hg, Pb, Se, and Zn all show an increased concentration in the smaller fly ash particles as compared to the bottom ash material.[12]

Cumulative frequency diagrams have been prepared for V, Ba, and Pb to illustrate differences in chemical concentrations for the ashes from three coal sources (Figure 15). Similarly, cumulative frequency diagrams for Se, Pb, and Cr (Figure 16) illustrate the difference in chemical concentrations for the three types of ashes.[12]

The chemical composition of the ashes is affected by the type of coal used in the power plant. Ba and Sr levels are the highest in ashes from western coals. Co and Hg levels are the highest in ashes from eastern coals. Cd, Mn, Mo, Ni, Pb and Zn are present in largest quantities in ashes from the midwestern coals. Concentrations of Cr, Cu, Se, V, and As are lowest in the ashes from western coal, but the concentrations of these elements is approximately the same for eastern and midwestern coal ashes.

Not only are these differences in the chemical composition of ashes due to coal source and the ash type, there are also variations in the composition of ashes due to time of sampling within the same power plant.[29,30]

Research has been conducted to determine what constitutes a representative utility waste sample, and to measure how variable the waste stream is over time.[25,29,30] The results show that reproducibility in chemical composition of solid wastes varies, as would be expected, for the waste type, elements measured, analytical techniques used, and the laboratories conducting the analyses.[25] Results show that the cause of greatest variance in reproducibility was attributable to the variability among laboratories analyzing the same aliquots of extract. With regard to sampling and time, the results indicate that the total variability in different degrees depends on the element and the method of analyses. Thus, the more samples collected and analyzed, the more precise the confidence level.[29,30]

2. Chemical Composition of Coal Ash Leachates

Several laboratory tests have been conducted to measure concentrations of dissolved chemicals in the aqueous phase. These tests use liquid extractants, usually water with some acid added to it, and a known amount of waste solids to generate the leachates under specified conditions of testing. The United States Environmental Protection Agency (USEPA) has specified a procedure called the EP Test (Extraction Procedure) to generate and measure leachate concentrations for the purpose of determining if a solid waste is hazardous. The American Society for Testing and Materials (ASTM) has also developed a standard leaching test known as ASTM Method A. Both of these methods are considered to be batch leaching procedures. The major difference between the USEPA and ASTM extraction procedures is that the USEPA test uses a weak acid and the ASTM test uses distilled water.

In contrast, a dynamic laboratory test consists of using a column of known size and length, filling it with the solid waste, and passing the extracting solution continuously through the column containing the waste material. The leachate is the solution collected after it has passed through the column.

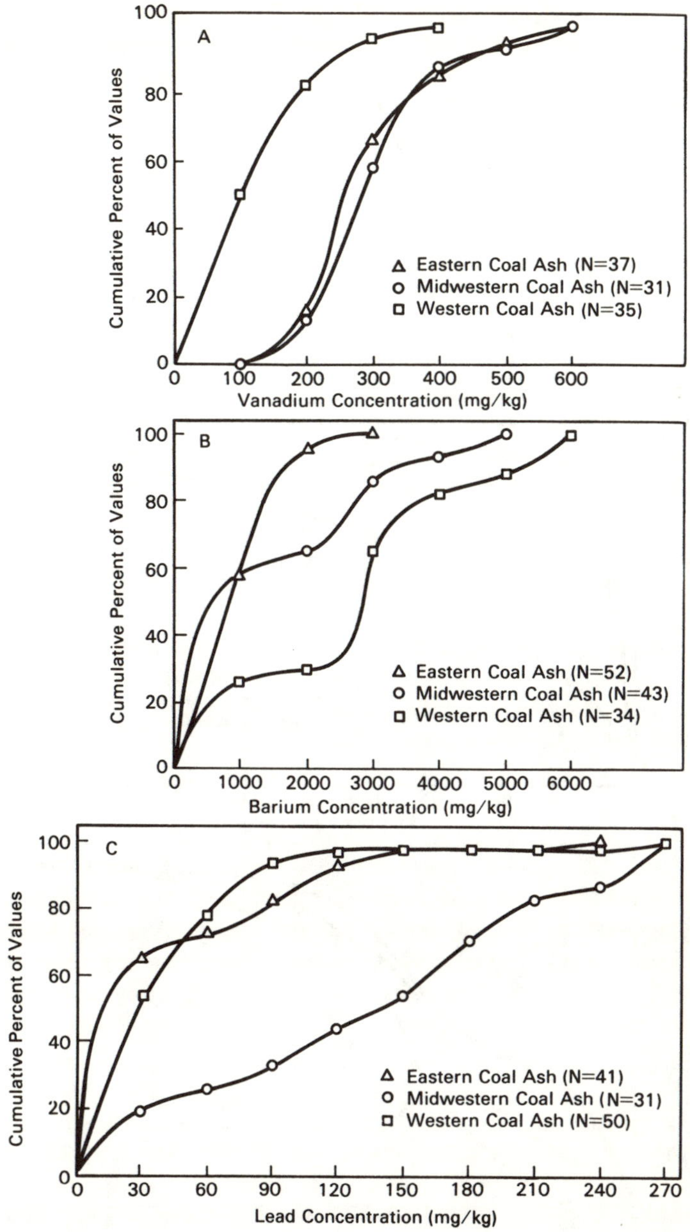

FIGURE 15. Cumulative frequency diagrams for V, Ba, and Pb in coal ashes from three coal types.[12]

In addition to the static batch and dynamic column tests, there is a "sequential extraction" method to generate leachates, particularly in a batch mode. In a sequential extraction procedure, after the agitation of solids and extracting liquid has been completed for the fixed time period, the solids and liquids are separated by vacuum filtration and a new extracting liquid is used with the solids. This procedure is repeated several times, resulting in a number of successive measurements indicating dissolvability of the solids.

All of these extraction, or leachate generation methods depend on a number of con-

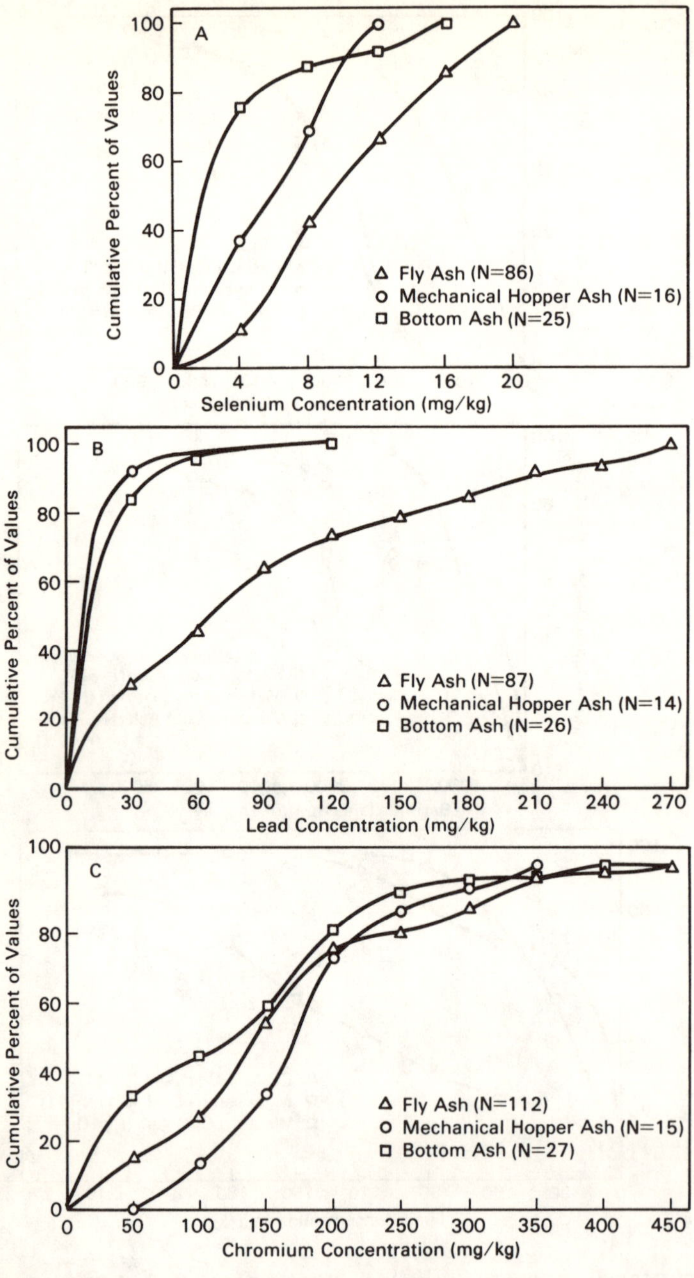

FIGURE 16. Cumulative frequency diagrams for Se, Pb and Cr in different types of coal ashes.[12]

ditions which constitute the testing procedures, among them: (1) solid to extracting liquid ratios, (2) the pH of the extracting solution, (3) the time period of agitation, (4) the hydrologic (flow rates) conditions, and (5) the temperature conditions.

The most commonly measured inorganic elements in the leachates from coal ash are those which are required by the USEPA in the use of EP test methods. Available data from the batch leaching tests are summarized in Table 12.[12]

Table 13 gives the leachate concentration data by coal ash type.[12] These tables show that both the ASTM and the EP method yield similar values for concentrations of Ag,

Table 12
TRACE ELEMENTS EXTRACTED FROM ASH IN SHORT-TERM BATCH TESTS[12]

Test	Ag	As	Ba	Cd	Cr	Hg	Pb	Se
EP	.00064	.012	.222	.0047	.036	.00042	.0050	.010
	<.0001—.20	.0004—1.46	.003—7.6	.0001—1.4	.001—0.68	<.0001—.007	>.0001—.25	<.0001—.17
	46	61	59	62	62	44	62	61
ASTM-A	.0007	.0072	.208	.00039	.047	.00027	.0025	.019
	<.0001—.03	.0005—.37	.0004—3.8	.0001—.013	.0005—.50	<.0001—.0012	<.0001—.25	.0005—.48
	24	25	24	25	25	19	25	25
Other extraction procedures[a]	.0032	.052	0.27	.0061	0.53	.00055	.033	.28
	.00025—.56	.0003—4.4	.013—40.	.0001—12.	.0005—210.	<.0001—0.15	.0001—1.6	.0003—32.4
	12	23	17	27	33	16	25	34

Note: Wastes include fly ash, mechanically collected ash and bottom ash, from eastern, western, and midwestern coals. Values given are: first line, geometric mean extraction concentration; 2nd line, range of values (mg/ℓ); 3rd line, number of extraction tests performed.

[a] Extraction times range from 2 to 48 hr. Final pH ranges from 3 to 11.5. Extractant-to-ash ratio ranges from 5 to 50.

Table 13
EP TEST RESULTS FOR COAL ASH, BY ASH TYPE AND COAL SOURCE[12]

Element	Bottom ash		Hopper ash		Fly ash	
	Eastern/midwestern	Western	Eastern/midwestern	Western	Eastern/midwestern	Western
Arsenic	2.2 ± 1.9 (4)	1.6 ± 4.8 (12)	51 ± 2.9 (4)	21 (1)	79 ± 4.4 (17)	9.6 ± 4.3 (22)
Barium	61 ± 6.4 (5)	144 ± 1.7 (12)	96 ± 1.1 (4)	187 (1)	134 ± 2.3 (16)	455 ± 2.3 (23)
Cadmium	8.5 ± 7.7 (3)	0.67 ± 5.3 (12)	24 ± 1.9 (4)	0.20 (1)	56 ± 3.4 (17)	2.4 ± 5.9 (23)
Chromium	18 ± 1.3 (4)	3.0 ± 4.2 (12)	120 ± 3.2 (4)	223 (1)	48 ± 5.3 (17)	104 ± 3.6 (23)
Lead	17.8 ± 13 (6)	1.9 ± 9.4 (13)	12.0 ± 4.2 (4)	15 (1)	19 ± 3.9 (17)	3.1 ± 5.5 (10)
Mercury	0.2 ± 3.5 (2)	1.1 ± 2.0 (9)	0.33 (3)	0.5 (1)	0.74 ± 3.1 (17)	1.3 ± 2.6 (23)
Selenium	5.2 ± 6.9 (4)	1.3 ± 2.3 (12)	15.9 ± 2.1 (4)	166 (1)	20 ± 2.1 (17)	23 ± 5.1 (22)
Silver	0.32 ± 1.9 (2)	0.48 ± 1.3 (11)	0.71 ± 2.0 (4)	0.30 (1)	1.2 ± 8.6 (4)	1.1 ± 4.8 (20)

Note: Values are geometric mean concentrations in $\mu g/\ell$ ± the standard deviation. Number of samples is given in parentheses. If coal source or ash type were not available, the values were not used to calculate these statistics. North Dakota lignite ash samples were included with ash derived from other western coals.

Table 14
FLY ASH BULK COMPOSITON AND LEACHATE COMPOSITION RESULTS[28]

	Solids concentration					
	Before leaching		After leaching	Leachate concentration		
Parameter	ICAP	NAA	NAA	Day 1	Day 3	Day 8
As	50.3	111	98.8	0.119	0.144	0.1615
Se	7.4	17.37	13.4	0.127	0.216	0.2225
Sb	—	9.25	8.1	0.124	0.151	0.168
Al	148,000	128,500	140,000	<0.01	<0.01	<0.01
B	61.1	—	—	6.24	6.49	6.33
Be	12.3	—	—	<0.0005	<0.0005	<0.0005
Cd	<1	<2	<14	<0.01	0.01	0.025
Co	36	58.5	54.9	<0.05	<0.05	<0.05
Cr	110	156.5	160	<0.01	0.01	0.03
Cu	167	—	<120	<0.008	<0.008	0.0085
Fe	44,000	46,500	44,120	0.03	0.08	0.085
Mn	188	204.5	192	0.15	0.1	0.09
Mo	<20	—	<310	1.6	1.9	3
Ni	106	155.5	<69	0.12	0.07	0.055
Sr	968	1,365	870	7.55	7.78	7.725
Ti	8,910	8,910.5	8,310	<0.005	<0.005	<0.005
V	269	272	295	<0.005	<0.005	0.0785
Pb	60	178	—	<0.05	<0.05	<0.05
Zn	109	—	<104	<0.05	<0.05	<0.05
Zr	226	622.5	725	<0.05	<0.05	<0.05
Ca	8,840	10,120	8,570	408	409	401.5
Mo	7,530	4,750	5,180	94.5	100	105.5
K	26,600	—	27,000	4,010	4,060	3,945
Na	4,100	2,565	3,000	53	54	51.5
pH	—	—	—	7.4	7.9	6.5
Conductivity	—	—	—	14,000	14,800	11,900
F	—	—	—	2.8	2.6	3.2
Cl	—	—	—	11.6	11.6	11.6
NO_3	—	—	—	6,770	6,390	6,280
SO_4	—	—	—	1,520	1,100	1,505

Note: All data are reported as parts per million (ppm). <, value indicates below detection limit.

As, Ba, Cd, Cr, Hg, Pb, and Se. The results also indicate that leachates from fly ash were higher in concentrations for As, Ba, Cd, Cr, and Se when compared to bottom ash or hopper ash. The leachates from eastern-coal ash are higher in As, Cd, and Pb content, whereas Ba content is higher in the leachates from western-coal ash.

Another study on leachate generation under laboratory conditions has recently been completed.[28] The results for one coal fly ash are given in Table 14. For the short-term batch leaching tests, several chemicals (Al, Be, Co, Pb, Ti, Zn, and Zr) were found to be below detection limits indicating essentially no leaching. There were several other elements (B, Ca, Mg, Na, Sr, SO_4, and Cl) which showed a very rapid release followed by slow or no further release indicating quick dissolution.

A relatively long-term (two years long) study to examine the leachability of fly ash using continuous flow-through columns was recently completed.[41,42] The results give a good understanding of the long-term changes in the chemical characteristics of fly ash leachate (Figure 17). One of the noticeable changes shown by these studies is the con-

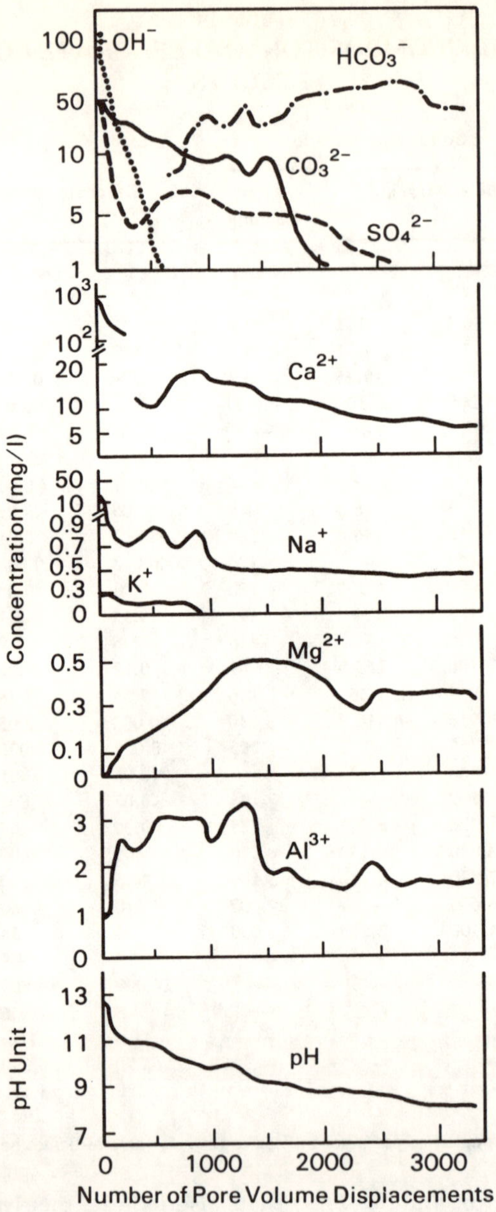

FIGURE 17. Leachate composition results from a laboratory column experiment using fly ash.[12]

tinuing decrease in measured pH of the leachate (from an initial value of 12.8 to a value of 7.5). This may be due to the weak acid forming gas CO_2 from the air entering the solution. Some mineralogical changes have been observed in the residual fly ash after the leaching experiment was finished. The measurements on the minor elements also show important leaching characteristics. Chromium showed an initial rapid release followed by a slow but stable rate of release. Cadmium and cobalt releases were at a constant rate throughout the experiment.

Based on the various batch and column leaching tests, it is clear that several of the chemicals present in the coal ashes are soluble and would dissolve when contacted with

Table 15
MAJOR COMPONENTS OF WET-FGD SLUDGE SOLIDS[12]

Coal source	Scrubber system	$CaSO_4$[a]	$CaSO_3 \cdot 1/2 H_2O$	$CaCO_3$	$Na_2SO_4 \cdot 7H_2O$	Inert[b] materials	Other
Eastern	Lime	19.4	69.2	10.3	—	—	—
		6.3	48.8	2.5	—	40.5	1.1 $MgSO_4 \cdot 6H_2O$
	Limestone	21.9	18.5	38.7	—	20.1	4.6 $MgSO_4 \cdot 6H_2O$
		15.4	21.4	20.2	—	40.9	3.7 $MgSO_4 \cdot 6H_2O$
		31.2	21.8	4.5	—	40.1	1.9 $MgSO_4 \cdot 6H_2O$
Eastern	Lime	15.1	37.4	21.5	—	15.9	—
Western	Lime	95	2	—	—	3	—
Eastern	Double alkali	15.3	68.1	10.1	3.7	—	—
		13.9	68.1	8.2	3.5	—	—
		5—25	65—90	2—10	—	—	—
Western	Limestone	84.6	8	6.3	—	3	1.5NaCl
Eastern	Double alkali	67.5	12.9	7.7	6.9	7.4	—
Western	Lime	17.3	10.8	2.5	—	58.7	10.7 "soluble salt"
	Alkaline fly ash	20	15	—	—	65	—
Eastern	Lime	19	12.9	0.2	—	59.7	8.2 "soluble salt"
Western	Double alkali	81.5	0.2	10.8	—	8.6	—
Eastern	Limestone	23	17	15	—	45	—
Eastern	Limestone	15	50	20	—	15	—
Eastern	Limestone	5	20	74	—	1	—
Eastern	Limestone	15	40	30	—	15	—

[a] Includes $CaSO_4 \cdot 2H_2O$, $CaSO_4 \cdot {}^1/_2H_2O$, and anhydrite.
[b] Chiefly fly ash.

water or other liquids. However, the dissolved solution concentrations are variable and are generally a small fraction (less than 1%) of the waste mass.

C. Chemical Characteristics of FGD Sludges

Since the primary purpose of FGD is to remove the gaseous form of sulfur by inducing reactions to convert the gas into solids, the chemical composition of the FGD sludges is necessarily dominated by the absorbing reagents, the sulfur content, and the entrained particles in the flue gas. Calcium or sodium-based salts are used in the wet scrubbers. Naturally occurring lime or limestone is used and both contain solid phases consisting of Si, Al, and Fe oxides as impurities. Reagents typically are used in excess quantities (10 to 30% excess) to assure maximum removal of sulfur gases.

Alkaline fly ash in a plant is sometimes used as the adsorbent. However, in most plants upstream ESPs are used to collect fly ash as a separate collection operation. The fly ash content of the FGD sludge, therefore, may vary from less than 10% of the solids (in plants which use particle collection systems prior to the FGD system) to over 50% of the solids (where particle collection systems are not used).

1. Bulk Composition

Only limited amounts of data are available on the chemical composition of FGD sludges (Table 15). These data show that calcium sulfates, calcium sulfites, and calcium carbonate are the principal chemical compounds present in the sludge. It is interesting to note that the lower-sulfur western coal produces scrubber sludge which is high in $CaSO_4$ content. In contrast, the scrubber sludge from eastern-coal plants tends to be higher in $CaSO_3$. A small amount of $MgSO_4$ is also present in the FGD sludges.

Table 16
TRACE ELEMENTS IN WET-FGD SLUDGES

	Sludge solids			Sludge liquor		
Element	Median conc. (mg/kg)	Range (mg/kg)	Number of values	Median conc. (mg/ℓ)	Range (mg/ℓ)	Number of values
Arsenic	12	0.8—52	11	0.03	<0.0004—0.06	17
Boron	140	42—530	4	14.9	2.1—76	8
Cadmium	10.6	0.06—25	12	0.02	0.002—0.12	17
Chromium	15	1.6—180	11	0.02	<0.0002—0.25	16
Copper	17.5	6—340	12	0.03	0.0045—0.53	16
Fluoride	625	266—1017	4	2.3	0.2—63	16
Mercury	0.4	0.005—6	9	0.005	0.00006—0.10	15
Lead	2.4	0.25—290	11	0.03	<0.005—0.50	17
Selenium	5.0	2—60	12	0.18	0.003—1.9	15

In addition to the major calcium compounds, there are a number of other inorganic elements which have been observed at smaller concentrations. Similar to fly ash and bottom ash, the concentrations of these chemicals vary in a large range. Only As, B, Cd, Cr, Cu, F, Hg, Pb, and Se data are available and summarized in Table 16. In a recent study, a dry FGD process sludge was analyzed for chemicals given in Table 17.[28]

2. Chemical Composition of FGD Sludge Leachates

Very little information is available on the leachability of FGD sludges. EP test results for three FGD systems are given in Table 18. Results from other short-term batch extraction methods are given in Table 19. Because of the dominance of sulfur compounds in the FGD wastes, the leachates show relatively high sulfate concentrations.

In a recently completed study on leachate generation, a dry-sorbent FGD waste was leached in laboratory batch tests.[28] The results are given in Table 20. For these short-term batch leaching tests, several chemical constituents (Al, Be, Co, Mn, Ti, Zr) were below detection limits when measured in the leachates, indicating very little or no leaching. There were several other inorganic elements (K, Cl, Cd, and So_4) which showed a rapid release followed by a slow or no further release indicating rapid dissolution. A few constituents (SF_4, Se, B, and Ca) showed an initial rapid increase in concentrations followed by decrease in concentrations in the leachate, indicating precipitation or sorption by the waste solids subsequent to dissolution.

Limited amounts of data on column tests with the FGD sludges are available and are shown in Figure 18.[43] One of the interesting observations which can be made from the study by Van der Sloot is that the pH of the column leachate rapidly decreased from 12 to about 9.5 and then stayed at 9.5 throughout the remaining 9 months of the experiment. Due to the presence of fly ash in the FGD sludge, the leaching characteristics of the sludges appear to be similar to the fly ash.

VI. CURRENT DISPOSAL METHODS

For high-volume wastes, approximately 75% of the material is disposed of and the remaining portion is reused. Landfilling (dry disposal) and ponding or impounding (wet disposal) are the most commonly used disposal methods. Occasionally, the coal ashes are disposed of in mines. During the 1950s and the 1960s, the disposal method almost always used was ponding; however, landfilling is increasing in the 1980s. The landfills and ponds often receive fly ash and bottom ash together, and a small fraction

Table 17
DRY FGD PROCESS SLUDGE BULK CHEMICAL COMPOSITION RESULTS[28]

Parameter	Concentration on dry weight basis (ppm) NAA	ICAP
Al	48,910	54,800
As	9.3	2.5
Ca	170,000	193,000
Cd	<8	2
Co	13	<5
Cr	67	66
Cu	<153	40.7
Fe	62,340	43,700
Mg	2,755	7,010
Mn	369	344
Mo	<11	<20
Na	5,102	6,800
Ni	39	123
Sb	1.6	—
Se	6.4	7.6
Sr	899	921
Ti	2,589	2,810
V	249	216
Zn	—	196
Zr	169	83

Note: These results are taken from EPRI Report No. EA-4215.

Table 18
RESULTS OF EP EXTRACTIONS OF FGD SOLID WASTES[12]

Extract Concentration (mg/ℓ)

Constituent	Lime sludge (midwest coal)	Lime sludge (lignite)	Ca-based dry scrubber waste Raw	Compacted
TDS	—	—	8650—8960	700—4590
Arsenic	0.06	0.14	0.001—1.6	<0.001—0.03
Barium	1.6	<0.40	—	—
Cadmium	0.015	<0.025	0.001—1.5	<0.001—0.008
Chromium	0.047	<0.10	—	—
Mercury	0.002	<0.001	—	—
Lead	0.016	<0.50	—	—
Selenium	0.031	0.11	0.018	—
Silver	—	<0.060	—	—
SO_4^{2-}	—	1300	600—610	98—190

Note: EP extraction is by distilled water; 1:16 solid/liquid ratio; 24-hr test.

Table 19
RESULTS OF VARIOUS SHORT-TERM BATCH EXTRACTIONS OF FGD SOLID WASTES[12]

Constituent	Wet lime/limestone wastes (acidic)[a]	Wet lime/limestone wastes (unbuffered)[b]	Dry Ca-based FGD wastes (unbuffered)[c]
Final pH	4.5—5.4 5.0—12.6	10.5—12.5	—
TDS	—	—	3230—9980
Arsenic	0.16—0.35	<0.002—0.15	0.005—0.04
Barium	0.073—0.4	<0.3—2	—
Cadmium	0.001—0.025	0.0005—<.025	0.015—0.031
Chromium	0.001—0.1	<0.001—0.1	0.1—1.5
Mercury	<0.001—0.001	0.0008—0.002	0.001
Lead	0.014	0.0027—0.0061	<0.001—0.01
Silver	0.001—<0.06	<0.06	—
SO_4^{2-}	1480—2260	1960	100—4200

[a] Two-day tests, pH buffered to 4 to 4.5, solid-liquid ratios of 1:4 and 1:16.
[b] One and two-day tests, extraction by distilled water, solid-liquid ratio of 1:4.
[c] Two and five-day tests extractions in distilled water, solid-liquid ratios ranging from 1:1 to 1:1000.

Table 20
DRY FGD PROCESS SLUDGE BULK COMPOSITION AND LEACHATE COMPOSITION RESULTS[28]

Parameters	Solid concentration: Before leaching	Solid concentration: After leaching	Leachate concentration: Day 1	Leachate concentration: Day 3	Leachate concentration: Day 7
As	12.15	8.1	BDL	BDL	0.00015
Se	15.7	7.8	0.0043	0.0015	0.00034
Sb	1.94	2.3	0.001	BDL	BDL
Al	46,950	390,000	<0.01	<0.01	<0.01
B	—	—	10.4	6	1.04
Be	—	—	<0.0005	<0.0005	<0.0005
Cd	—	<8	0.02	0.04	0.04
Co	13.45	11	<0.05	<0.05	<0.05
Cr	78	62	0.65	1.08	1.54
Cu	<3,300	<100	<0.008	<0.008	0.013
Fe	48,150	40,000	0.04	0.08	0.085
Mn	375	295	<0.01	<0.01	<0.01
Mo	—	<11	1.4	3.2	3.65
Ni	209	<40	<0.05	<0.05	0.095
Sr	1,790	890	13.8	19.7	28.1
Ti	2,970.5	2,270	<0.005	<0.005	<0.005
V	238	206	<0.005	<0.005	0.016
Pb	62	—	<0.05	<0.05	—
Zn	448	34	<0.05	<0.05	0.075
Zr	210	172	<0.05	<0.05	<0.05
Ca	209,500	160,000	1,560	1,460	1,040
Mg	2,715.5	2,100	<0.01	<0.01	0.105
K	—	8,500	3,860	3,680	3,980
Na	6,590	4,500	72	82	101.5
pH	—	—	12.3	12.3	12.6
Conductivity	—	—	19,700	>20,000	19,500
F	—	—	5.8	4.6	3.2

Table 20 (continued)
DRY FGD PROCESS SLUDGE BULK COMPOSITION AND LEACHATE COMPOSITION RESULTS[28]

	Solid concentration		Leachate concentration		
Parameters	Before leaching	After leaching	Day 1	Day 3	Day 7
Cl	—	—	225	245	245
NO_3	—	—	5,550	5,700	5,700
SO_4	—	—	1,540	1,540	690
Alkalinity	—	—	43.6	42.8	43.1
SO_3	—	—	29	5	24.5

Note: All data are reported as parts per million (ppm). BDL, below detection limit.

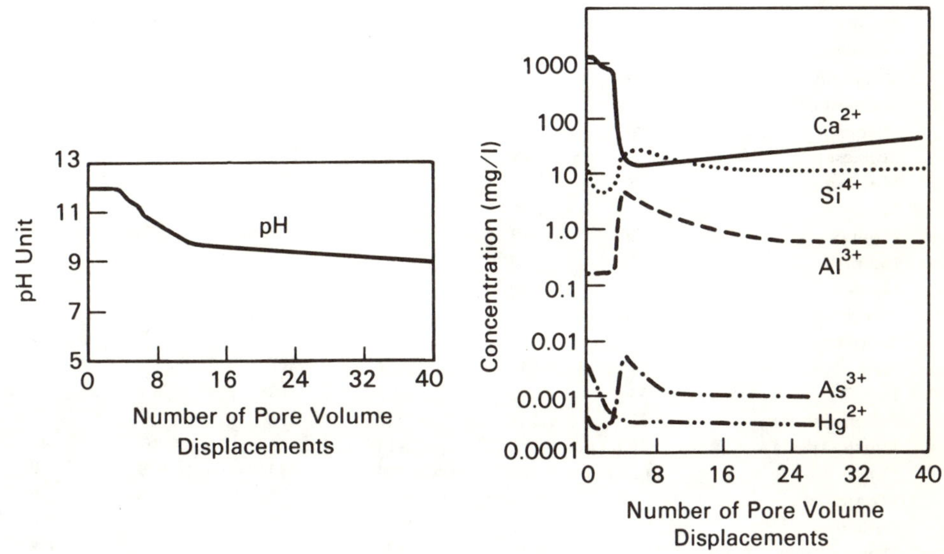

FIGURE 18. Leachate composition results from a laboratory column experiment using a Ca-based spray dryer FGD waste.[12]

of ponds and landfills receive bottom ash, fly ash, and sludge altogether. Because of the collection which combines fly ash and the scrubber waste material, there are landfills and ponds which receive this combined waste. Table 21 provides the summary statistics of quantities disposed of by state for the year 1981.[23]

In a recent survey of 64 plants with a total capacity of 50,900 MW, information on ash disposal practices was developed.[44] Most of the plants in the survey began operation in 1970 to 1978. and hence, are relatively new. The data were characterized according to the disposal method used for each type of solid waste, the total coal-fired generating capacity of the plant, the solid wastes collection equipment, and the type of FGD systems used. The data obtained are summarized in Table 22 for fly ash. From the survey, it appears that:

1. The greater number of plants use ponding of fly ash.
2. The majority of plants use ESP to collect dry fly ash, but dispose of the ash by ponding.

Table 21
SOLID WASTES DISPOSAL SYSTEMS AND QUANTITIES FOR 1981, BY STATE[23]

	Fly ash (1000 tons)		Bottom ash (1000 tons)		FGD sludge (1000 tons)	
State name	Landfilled	Ponded	Landfilled	Ponded	Landfilled	Ponded
Alabama	564	1474	195	251	0	0
Arkansas	172	0	84	0	0	0
Arizona	638	460	213	131	0	21
Colorado	315	0	59	0	0	0
Delaware	214	0	1	0	0	0
Florida	31	288	0	109	0	0
Georgia	0	1,679	0	336	0	0
Illinois	228	796	231	666	181	142
Indiana	165	1,817	53	656	49	1
Iowa	106	384	18	113	0	0
Kentucky	531	1,309	46	385	97	63
Kansas	377	19	492	4	392	0
Louisiana	0	1	0	10	0	0
Maryland	558	33	156	8	0	0
Massachusetts	158	0	14	0	0	0
Michigan	167	964	36	150	0	0
Minnesota	27	648	12	150	0	54
Mississippi	68	178	8	49	26	0
Missouri	124	996	91	473	19	0
Montana	320	0	52	0	0	0
Nebraska	145	0	51	0	0	0
Nevada	354	0	166	0	0	0
New Hampshire	6	0	0	0	0	0
New Jersey	1	0	0	0	0	0
New Mexico	647	1,181	152	297	0	0
New York	334	0	51	0	0	0
North Carolina	814	1,364	187	301	0	0
North Dakota	180	112	114	254	67	47
Ohio	1,051	3,132	204	934	680	0
Oklahoma	0	0	0	31	0	0
Oregon	29	0	21	0	0	0
Pennsylvania	2,391	1,715	435	207	0	328
South Carolina	95	382	19	101	0	12
South Dakota	55	0	128	0	0	0
Tennessee	0	1,784	0	467	0	0
Texas	2,525	85	1,029	65	516	0
Utah	0	0	0	0	0	0
Virginia	452	0	93	34	1	0
Washington	0	0	0	0	0	0
West Virginia	1,839	566	238	276	0	0
Wisconsin	306	228	69	79	0	0
Wyoming	726	0	455	45	61	0
U.S. total	16,719	21,598	5,175	6,581	2,099	1,288

3. Nearly 17% of the plants pay another company to dispose of or otherwise haul away the ash, and 5% sell the ash directly to some company.

The above-mentioned survey also included data on bottom ash disposal which is summarized in Table 23.[44] It appears that:

1. On a plant basis, the vast majority use water to remove bottom ash from the plants, but less than half of these plants use ponding for disposal.

Table 22
FLY ASH COLLECTION AND DISPOSAL PRACTICES[44]

Method	Number of plants reporting method	Percent of plants reporting method
Collection		
Dry electrostatic precipitator	39	61
Mechanical (baghouse, etc.)	8	13
Wet electrostatic precipitator	4	6
Particulate scrubber	2	3
Other	11	17
Total	64	100
Disposal		
Ash pond	26	40
Conveyed to landfill (dry)	19	30
Paid disposal	11	17
Sale of fly ash	3	5
Intermediate ponding followed by landfill	3	5
Other	2	3
Total	64	100

Table 23
BOTTOM ASH COLLECTION AND DISPOSAL PRACTICES[44]

Method	Number of plants reporting method	Percent of plants reporting method
Collection		
Wet sluiced	52	81
Dry conveyor	11	17
Other	1	2
Total	64	100
Disposal		
Ash pond	24	38
Conveyed to landfill (Dry)	17	27
Paid disposal	8	12
Sale of bottom ash	6	9
Intermediate ponding followed by landfill	6	9
Other	3	5
Total	64	100

2. Mechanical dewatering and dry disposal of bottom ash do not appear to be common practice.
3. On a quantity basis, the most common disposal method is ponding.

In the case of wet scrubbers, wastes from particulate control, SO_2 control, or combined particulate and SO_2 control systems, there are two basic approaches; the wastes can be pumped directly from the scrubber to the disposal pond, or the wastes can be dewatered first via thickening/clarification prior to being pumped or trucked to the pond. In some wet-scrubber systems for SO_2 control, the wastes (thickened or unthickened) are first mixed with sluiced ash or combined with additives for stabilization.

In almost all cases, provision is made for returning supernatant to the process. This is particularly true for ash ponding systems and FGD systems discharging unthickened wastes directly to ponds. Here, water management and conservation dictate water reuse. However, not all ponding systems incorporate water recycle, and many which do include this feature, practice only partial recycle.

A. Disposal in Ponds/Surface Impoundments

The power plants which handle their solid wastes in a slurried or wet form usually dispose of the material in ponds or surface impoundments. These ponds may be unlined or lined by natural or artificial materials. Both fly ash and bottom ash are often combined and sluiced to the ponds which are located on the power plant site. Examples of the wet disposal ponds are in Figures 19 to 22. The common factor in all wet disposal systems is that the solids are transported and deposited in the disposal site by water.

Once the wastes are sluiced to the ponds, the solid particles tend to settle at the bottom of the pond. Often, these ponds are dewatered and the settled solids are removed to landfills or sold for reuse. In 1981 about 56% of the disposal of fly ash and bottom ash was in ponds. Approximately 40% of the FGD sludge was disposed of in ponds. The size of ponds varies in a wide range between 25 acres to over 350 acres each. Average size of a pond is 200 acres with a life of about 15 years. Listed in Table 24 are the types and number of ponds used in the electric utility industry in 1981.[23]

B. Disposal in Landfills

Dry disposal methods may involve any one of the following modes:

1. Dry collection and direct disposal of coal ash and dry sorbent FGD wastes
2. Interim ponding followed by excavation and landfilling
3. Mechanical dewatering and direct landfilling of FGD wastes
4. Blending with fly ash and direct landfilling of FGD wastes
5. Stabilization through the use of additives (nonproprietary or otherwise) prior to landfilling

The exact nature of a particular operation will depend primarily upon the type of waste and the location of the disposal site. Where disposal sites are at some distance from the power plant (or waste processing plant), transport of dry, dewatered, or processed waste is usually accomplished by open, rear-dump trucks although in at least one case, a dedicated rail-haul system is being used. Coordinating a trucking operation with waste production and disposal frequently requires waste transfer/storage systems, especially at the power plant (or processing plant). In the simplest cases this can involve directly filling the trucks with wastes via feeder chutes or hopper bins (although provisions for temporary storage areas are required). This approach frequently is used in dry ash disposal and some codisposal operations. In some cases, though, particularly for FGD/ash codisposal and disposal of stabilized wastes, it is necessary to provide interim storage piles from which the waste is placed in trucks with front-end loaders. This usually requires use of stacking conveyors.

Where the disposal site is adjacent to the power plant, the trucking of the wastes can be minimized or possibly eliminated by locating the waste dewatering or processing plant at the disposal site. Stacking conveyors can then be used to move the wastes into the disposal areas, and earth-moving equipment is used to spread and compact the wastes. This is usually practical only for relatively small disposal operations.

In general, operations at the landfill area involve dumping, spreading, and compaction. Usually, only small sections are worked at any one time. The wastes are layered in one- to three-foot lifts, with spreading and compaction most commonly accomplished using dozers. In some cases, fly ash (and bottom ash) and FGD waste can be codisposed without prior blending. In such operations, the ash and FGD wastes would

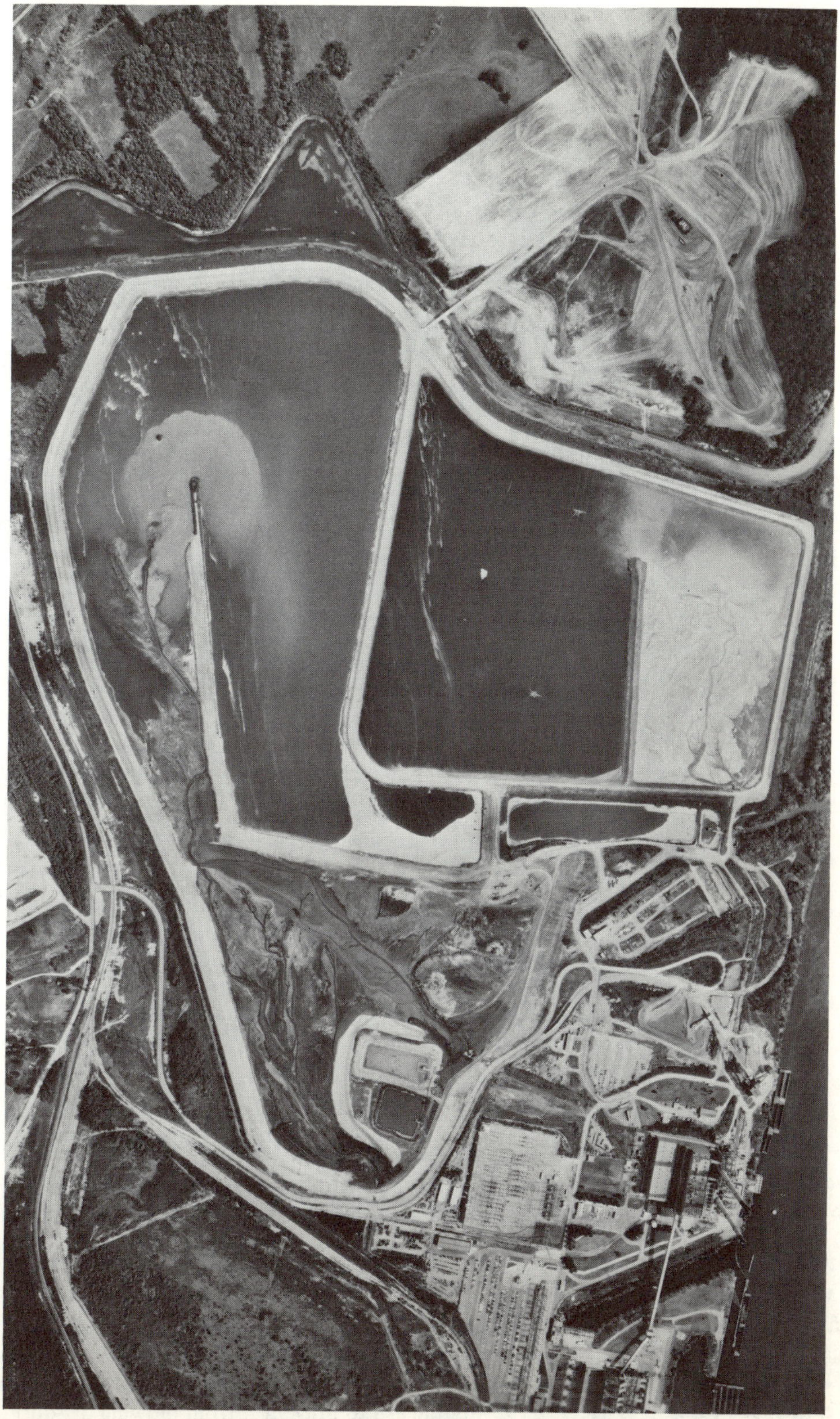

FIGURE 19. Aerial view of a power plant site showing solid waste disposal ponds.

FIGURE 20. Photographs showing coal ash disposal pond.

be placed in alternate layers. Listed in Table 25 are the types and numbers of landfills used in the electric utility industry in 1981.[23] Examples of dry disposal landfills are given in Figures 23 to 26.

1. Dry Disposal of Ash

Coal ash (fly ash and, if necessary, bottom ash) can be handled in a dry state. In such a case, the ash is conveyed pneumatically from the electrostatic precipitator or bag filter to storage silos for intermittent transfer for disposal.

Fly ash normally is deposited in trucks or railroad cars for transport to a disposal area. In such cases, it is necessary to wet the ash to prevent it from blowing off during transportation.

FIGURE 21. Photographs showing disposal ponds receiving fly ash and bottom ash.

The wetted coal ash is deposited in the fill site and spread by dozers in small lifts of about one to three feet; it is then compacted. When and where needed, water trucks are employed during disposal operations at the site to control fugitive dust. The lifts are built up to a total ultimate fill height which will be site-specific but which may range from about 30 feet to over 80 feet.

2. Interim Ponding

In this approach, the waste is settled in storage ponds and the supernatant is recycled to the process, or otherwise lost through evaporation. The pond may be reclaimed or, when the material is a moist solid, may be evacuated and landfilled in a permanent site. This method has been used for disposal of sluiced ash from dry collectors, and in at least some cases, for disposal of wastes from wet particulate scrubbing systems. The interim ponding/excavation method is most applicable in arid regions.

3. Mechanical Dewatering and Landfilling

Mechanical dewatering methods, including vacuum filtration or centrifugation, may

FIGURE 22. Photographs showing disposal ponds receiving FGD sludges and coal ashes.

remove enough water wastes to allow direct landfill disposal without additional processing, e.g., ash blending or stabilization. This method of disposal is applicable to:

1. Sluiced fly ash from dry collection systems where the ash is dewatered in settling/decanting bins
2. FGD wastes from systems producing sulfate-rich wastes
3. FGD wastes from systems employing forced oxidation to produce gypsum

4. Codisposal of Ash and FGD Wastes

In an engineering context, it may be easier to handle the sludge-ash mix than to handle (transport, place, compact, etc.) FGD waste or ash separately. Fill construction would be easier using one material rather than two at the same site (mixed wastes vs. layers of waste between ash), or using one material in one fill rather than two separate fills, at the same site or at separate sites. The mix of ash and sludge should be less compressible and may be self-hardening, or it may require less fixative in a mixed fill than in separate fills.

A possible disadvantage to this method could be an increase in permeability (ash-waste mix vs. waste alone) with potential for more infiltration and leaching. Another

Table 24
TYPES AND NUMBERS OF PONDS IN THE ELECTRIC INDUSTRY IN 1981[23]

	No. of fly ash	No. of bottom ash	No. of fly ash and bottom ash	No. of FGD sludge
Unlined ponds	58	40	64	5
Lined ponds				
Natural material	10	2	9	10
Artificial material	0	0	2	25

Table 25
TYPES AND NUMBERS OF LANDFILLS IN THE ELECTRIC INDUSTRY IN 1981[23]

	No. of fly ash	No. of bottom ash	Fly ash and bottom ash	No. of FGD sludge
Unlined landfills	42	20	25	15
Lined landfills				
Natural material	10	4	12	5
Artificial material	0	0	0	3

FIGURE 23. Photograph showing an active landfill for high volume solid wastes.

disadvantage is that the mix could be more difficult to transport than is ash alone (but easier than waste alone, at the same solids content).

5. Stabilization Additives

In this method of dry disposal, lime and fly ash or other stabilization additives are added to dewater wastes prior to disposal.

In a stabilization process, cementitious additives produce pozzolanic reactions that result in increased structural integrity in the wastes. This chemical process continues after placement. Typically, for dry impoundment type of disposal, the wastes are thickened and dewatered to a high solids content and blended with fly ash and lime, thus forming a material with cementitious properties.

FIGURE 24. A partially completed landfill containing fixed FGD sludge and ashes.

FIGURE 25. A partially completed landfill disposal site for fly ash and bottom ash.

The degree of processing necessary to allow dry disposal varies considerably, but usually involves thickening and dewatering processes. In scrubbing systems with a large amount of ash (e.g., alkaline-ash scrubbers and scrubbers without upstream particulate removal), extended settling in a pond may dewater the sludge sufficiently to allow excavation and landfilling. Other systems (e.g., those using high-sulfur, low-ash coals) may require several dewatering steps followed by the addition of chemical bonding agents (such as dry fly ash and/or lime) prior to disposal. Most dry systems include both dewatering and blending processes to produce a material suitable for landfill disposal.

Blending dry material with the sludge aids considerably in reducing its fluid nature. The most commonly used dry material is fly ash, because it is readily available at the station. As a result, dry disposal may be difficult or uneconomical for systems with little or no dry fly ash, i.e., utilities with cyclone-fired boilers, wet-ash handling or disposal systems, no upstream particulate removal, or marketable fly ash. Dry disposal systems are usually more applicable in arid regions, on systems with high ash/sludge ratios, and/or where land is at a premium.

The principal advantage of dry disposal is a more efficient use of the disposal site,

FIGURE 26. A landfilling operation for coal ashes.

thereby requiring less land area overall. The waste materials are generally more dense, do not require land area for dams and dikes, and can be placed in configurations not possible with wet disposal systems. When compared to those for wet systems, dry disposal sites are generally easier to develop and reclaim.

Landfills may also be unlined or lined by natural or artificial materials. Only a few landfills currently use liners composed of natural materials. Landfills require smaller land mass for disposal than ponds. Approximately 90,000 dry tons of utility wastes can be placed in one acre of landfill.

Once the landfills have been filled, they are covered with clay caps overlain by soils for reclamation of the land.

VII. REUSE OF COAL FLY ASH, BOTTOM ASH, AND SLAG

The electric utility industry has developed reuse markets for fly ash, bottom ash, and slag in a wide range of applications. Historically, on a national basis, reuse has varied between 15 to 25% of the tonnage generated.[22] For example, 18.2%, or 12.8 million tons of fly ash, bottom ash, and boiler slag were reused in various applications in 1983 (Table 26). While generation of fly ash, bottom ash, and slag has increased from about 39 million tons in 1970 to roughly 70 million tons in 1983 (Table 6), the reuse of these

Table 26
SOLID WASTE UTILIZATION FOR THE ELECTRIC UTILITY INDUSTRY — ANNUAL STATISTICS (1970 TO 1983)[22]

	Fly ash		Bottom ash		Boiler slag	
Year	Amount	% of Production	Amount	% of Production	Amount	% of Production
1966	1.4	7.9	1.7	21.0	0	0
1970	2.2	8.1	1.8	18.6	1.1	39.1
1971	3.3	11.7	1.6	16.0	3.7	75.2
1972	3.6	11.4	2.6	24.3	1.3	35.3
1973	3.9	11.4	2.3	21.9	1.8	44.3
1974	3.4	8.4	2.9	20.3	2.4	50.0
1975	4.5	10.6	3.5	26.7	1.8	40.0
1976	5.7	13.3	4.5	31.5	2.2	45.8
1977	6.3	13.0	4.6	32.6	3.1	60.0
1978	8.4	17.4	5.0	34.0	3.0	58.8
1979	10.0	17.4	3.3	26.4	2.4	46.0
1980	6.4	13.3	4.3	29.5	1.8	48.1
1981	9.4	19.0	4.1	32.0	2.9	57.0
1982	8.0	16.6	3.6	27.6	2.0	45.1
1983	7.5	14.3	2.8	19.7	2.5	64.2

Note: Values given in millions of tons.

wastes has increased from 5 million tons in 1970 to 12.8 million tons in 1983 (Table 26). The annual tonnages of coal combustion wastes reused since 1970 are given in Table 26. It should be noted that in terms of tonnage, fly ash is reused in the largest quantities, followed by bottom ash and boiler slag. As a percentage of quantities produced, however, the boiler slag shows reuse exceeding 50%, followed by bottom ash reuse of about 25%, with fly ash reuse of about 15%.[22]

There are several reuses for coal combustion wastes exhibiting various stages of development for each reuse. Some reuses, such as in cement and concrete, highway construction, and roofing granules, are at commercial scale. The degree of development for additional reuses ranges from bench-scale pilot studies (e.g., aluminum extraction, mineral wool production) to full-scale demonstration projects (e.g., artificial reefs). Table 27 lists known reuses for the high volume coal combustion wastes, showing commercially accepted uses as well as uses in the development stage. Selected key reuses are discussed below.

A. Cement and Concrete Markets

Coal ash as a raw material and/or pozzolan in cement and concrete has been recognized in the U.S. since the late 1930s. In 1937, R. E. Davis prepared a report which concluded fly ash could replace 20 to 50% of the Portland cement in a given mix because of its pozzolanic characteristics. Today, use in blended cement and concrete constitutes the largest demand for fly ash, about 3.6 million tons per year. The use of coal ash in cement and concrete has benefited from the USEPA guidelines addressing procurement and use of ash in these applications.[45] As described in Figure 27, coal ash can be introduced at any of four points in the process: (1) as a raw material in feed stock, (2) as a pozzolan interground with the clinker, (3) as a pozzolan mixed after the final grinding, and (4) as a cement replacement in the concrete manufacturing process. In 1983, 12.8 million tons of coal ash were utilized in the U.S. Of the 9.0 million tons utilized in commercial applications, 1.1 million tons were used in blended cement, and

Table 27
REUSES FOR COAL COMBUSTION BY-PRODUCTS

Reuse	Development status: Commercial	Development status: Potential[a]
Replacement for cement in concrete	X	
Raw material in brick manufacture		X
Cement replacement in precast concrete products	X	
Raw material ingredient in the manufacture of Portland cement	X	
Raw material for making mineral wool insulation		X
Aggregate for the stabilization of highway subgrades	X	
Lightweight aggregate for concrete		X
Controlled density fill for backfill of various construction	X	
Raw material for metal recovery		X
Fighting mine fires and controlling subsidence	X	
Soil amendment		X
Mineral filler in asphaltic concrete	X	
Plastic fillers	X	
Ingredient in grouting	X	
Oil well cementing	X	
Land reclamation		X
Blasting grit	X	
Deicing and abrasive agent	X	
Stabilization of waste ponds	X	
Coating for coal storage piles	X	
Feed in sulfur control (FGD systems)	X	
Structural fill for industrial parks	X	

[a] Note: This includes conceptual studies, pilot studies, laboratory studies, and demonstration projects.

just over 2.6 million tons were used as partial replacement of cement in concrete and concrete products.

When used as a component of the raw material feed stock in the production of Portland cement, fly ash serves as a source of silica, alumina, and iron. Also, the residual carbon contained in the fly ash enhances the firing process, adding energy to the kiln. The ash may also contain undesirable elements — sulfates, and alkalies such as potassium and sodium oxide — whose quantities must be controlled in a manner similar to other standard feed stock materials to produce satisfactory cements.

1. Fly Ash as an Ingredient of Blended Cements

Fly ash has traditionally been used in Europe, Japan, and to some extent, in the U.S., in the production of hydraulic cement. Recently, its use in hydraulic cements has had growing acceptance in the U.S., where it has grown from 0.16 million tons in 1970 to the 1.1 million tons in 1983. Several U.S. cement manufacturers produce a Portland-pozzolan cement containing fly ash, either through the blending of the ash with finished Portland cement, or through the intergrinding of ash with the clinker. This Type IP, or Type I-PM cement is suitable for use in general concrete construction. Benefits associated with its use include increased grinding efficiency, decreased cost of cement, improvement in quality of cement in terms of workability of concrete, reduced drying shrinkage, improved textural strength with prolonged curing, and improved chemical resistance of concrete in service.[45-47] As defined in ASTM Specification C 595, Type IP is permitted to have a fly ash content of 15 to 40% by weight, while Type I-PM must contain less than 15% by weight.

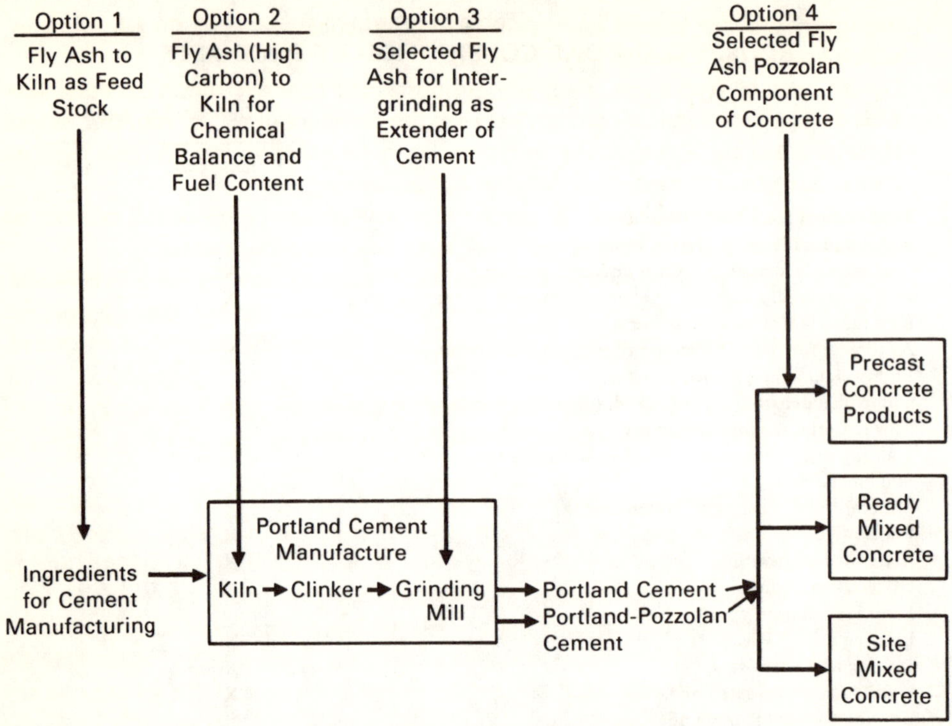

FIGURE 27. Schematic showing uses of fly ash in cement and concrete.

2. Fly Ash as a Component of Portland Cement Concrete

Fly ash can be used as a component of Portland cement concrete, either as an admixture with the cement, or as a replacement for a percentage of the Portland cement up to a maximum of 50%, depending on performance requirements. Fly ash use in Portland cement concrete production represents one of the major uses of fly ash at over 2.6 million tons in 1982.

When fly ash is used as an admixture in concrete, the concrete must exhibit early strength and durability equal to that specified for concrete not containing a finely divided admixture. Fly ash used to replace a portion of the cement, generally by the ready-mix concrete company, must be of an appropriate fineness and composition, and generally results in slower setting time, and lower early strength concrete, depending on how the mixture is proportioned. The slower setting time and lower early strength result in part from the composition and fineness of the particular ash employed, and in part from the proportions and properties of the other materials used.[46] Typical replacement rates are from 15 to 25% of the cement in a concrete mix.

The proper use of fly ash in concrete has been found to have many advantages. These advantages can be summarized as follows:[45]

- Greater ultimate strength. Proper proportioning with fly ash can produce concrete that can perform as well as or better than concrete made without fly ash, even under adverse curing conditions. Proper mix design achieves 28-day strength equal to or better than a typical Type I cement; 56-day strength is also higher in almost all cases.
- Improved workability. Fly ash used in properly proportioned concrete mixes improves the pumpability, handling, placing, and finishing of fresh concrete. These

improvements are in part due to the glassy, spherical shape of fly ash particles, which add to the "plasticity" of the mix.

- Lower heat of hydration. Properly proportioned concrete mixes generate less heat during the chemical reaction between the cement and water, resulting in less thermal cracking — especially important in mass concrete applications such as dams, large beams, retaining walls, and foundations.
- Increased sulfate resistance. A proper mix will reduce the chemical attacks on concrete from sulfates contained in adjacent soil and groundwater.
- Increased resistance to freezing and thawing. Fly ash reduces the voids in concrete — by mixing with certain cement constituents so they are not leached out, by filling minute voids, and by maintaining a more even distribution of entrained air.

Fly ash is variable in its physical and chemical composition, requiring adequate testing and quality control of the ash as provided for in ASTM specifications C 595, C 311, and C 618.

Due to the lack of widely publicized and sometimes inadequate data on mix designs (as compared to general purpose concrete), fly ash concrete may be improperly prepared by an unfamiliar user. Mix designs do exist to produce fly ash concrete, and in most cases, fly ash brokers/companies will provide manufacturing assistance and engineering design aid. Mix proportions have been developed which will achieve equal or superior strength to general purpose concrete at all ages, particularly if 10 to 30% fly ash is added to a concrete mix, or if a blended cement with up to 10% fly ash content is used.

The use of fly ash in cement and concrete is well documented in the U.S. and other countries. Federal agencies in the U.S., such as the Army Corps of Engineers and the Bureau of Reclamation, use fly ash extensively. In the private sector, the use of fly ash in ready-mix concrete is extensive, and is growing. The use of fly ash in cement manufacturing is also on the increase, but has not yet reached the level found in other countries.

3. Fly Ash in Concrete Blocks

Some manufacturers of blocks have found that fly ash can be as much as 50% of total cementing material in high-pressure, steam-cured block. Many nonautoclavers use from 20 to 35% fly ash in their mixes. The producers find that they can get the strength they require with improved finish and texture, better mold life, and better corners on the finished block, when they use fly ash. It appears that fly ash gives added plasticity to the relatively harsh mixes used in block.[46]

4. Fly Ash Concrete in Pipes

The use of fly ash in a pipe mix is particularly advantageous to concrete pipe manufacturers. Fly ash in a properly proportioned pipe mix increases the density and makes a much more watertight concrete. Another important factor is that concrete made with fly ash is more resistant to the weak acids sometimes encountered in sewer lines, and likewise resists the attack of sulfates.[46]

5. Fly Ash Concrete in Dams

The use of fly ash concrete in dams is well documented by the U.S. Bureau of Reclamation and by the Army Corps of Engineers. Fly ash concrete in the construction of dams has been used since Hungry Horse Dam in 1953. Some of the more recent dams constructed by the Army Corps of Engineers utilizing fly ash concrete include: Richard B. Russell Dam, Dworshak Dam, Libby Dam, and the Tombigbee Riverway.[64]

6. Fly Ash Concrete in Waste Treatment Plants

Fly ash concrete has been used in several waste treatment plants, for example: (1) sanitary wastewater treatment plants in Hammond, Ill.; Sycamore, Ill.; Oak Harbor, Ohio; and Allegan, Mich.; (2) Industrial wastewater treatments plants in East St. Louis, Ill.; Ladysmith, Wis.; and Park Fall, Wis.;[47] and (3) waste treatment plants in Cobb County, Ga.; and Atlanta, Ga. According to the report of ACI Committee 350, *Concrete Sanitary Engineering Structures,* Portland pozzolan cement ASTM C 595 can be used for the cement, but the pozzolan content should not exceed 25% by weight.

7. Fly Ash Concrete in Structural Buildings

Fly ash concrete is extensively used in buildings in Chicago, e.g., the Inland Steel Building, First Wisconsin National Bank of Milwaukee, John Hancock Center, Sears Tower McCormick Place, First National Bank Building, and the Prudential Building. Depending on its intended use, the percentage of fly ash in the mix ranges from 10 to 28% of the cement by weight. Fly ash concrete has also been used in the Southeast. Examples of buildings built with fly ash concrete include the C&S Bank Tower and the Atlanta Hilton in Atlanta, Ga., and a hospital, presently under construction, in Birmingham, Ala.[48]

B. Soil Applications

The physical and chemical properties of coal ashes have prompted scientific investigation into their possible use in agriculture and land reclamation. Characteristics such as silty or fine sandy texture, alkalinity, and abundance of plant nutrients including boron, sulfur, molybdenum, and zinc, have prompted researchers to look closely at fly ash and the possible advantages of its use in soil applications.[49]

1. Fly Ash as Source of Nutrients

All elements essential for growth which plants obtain from soil are marketed in some form of commercial fertilizer. The effectiveness with which fertilizers provide nutrients in a form available to plants depends on many factors. Elemental composition, the chemical form of nutrients and their solubility, pH (acidity or alkalinity) of soil and soil amendment mixture, and plant species being managed are all important. Except for nitrogen, these same essential elements occur in coal fly ash at concentrations comparable to or greater than those found in soil (Table 28 and 29). Fly ash contains sufficient concentrations of boron, sulfur, molybdenum (lignitic coal ash), and zinc (bituminous coal ash) to correct soil deficiencies in these elements under experimental conditions.[49-56] However, the concentration of a particular element in ash is not the only factor affecting its uptake by plants. Thus, changes in crop yields from ash-amended soils should not be attributed solely to a specific element being contained in the ash.

Much of the research on ash application to soils for plant nutrition shows that plant uptake is consistently increased by ash constituents boron,[50-51] sulfur,[49,56] and molybdenum.[53,55] Alfalfa and sweet clover require large amounts of sulfur. Modern nitrogen and double super-phosphate fertilizers, however, contain very little sulfur. It is commonly postulated that the switch to these high-analysis fertilizers is partly responsible for the current trend of increasing incidences of sulfur deficiency.[57,58] As crop yields and use of these high-analysis fertilizers increase, sulfur deficiency may become more widespread and the need for sulfur fertilizer more pronounced.

Fly ash use in agriculture does have possible problems restricting its widespread use. Since fly ash contains metals, soluble salts, and alkalinity, fly ashes can produce detrimental effects on vegetation. The alkalinity of fly ashes can be beneficial in humid

Table 28
ANALYSES OF EIGHT REPRESENTATIVE U.S. SURFACE SOILS[57] AND FLY ASHES

Constituents	Norfolk fine sand, Florida	Sassafras sandy loam, Virginia	Ontario loam, New York	Loam from ely, Nevada	Hagerstown silt loam, Tennessee	Cascade sislt loam, Oregon	Marhall silt loam, Iowa	Summit clay, Kansas	Fly ashes from Bit. coal	Fly ashes from Subbit. coal	Fly ashes from Lignite coal
SiO_2	91.49	85.96	76.54	61.69	73.11	70.40	72.63	71.60	48.76	49.69	30.69
TiO_2	0.50	0.59	0.64	0.47	1.05	1.08	0.63	0.81	1.45	1.09	0.74
Fe_2O_3	1.75	1.74	3.43	3.87	6.12	3.90	3.14	3.56	16.44	6.48	8.87
Al_2O_3	4.51	6.26	9.38	13.77	8.30	13.14	12.03	11.45	23.26	23.04	15.48
MnO	0.007	0.04	0.08	0.12	0.44	0.07	0.10	0.06	—	—	—
CaO	0.01	0.40	0.80	5.48	0.37	1.78	0.79	0.97	2.88	13.81	22.29
MgO	0.02	0.36	0.75	2.60	0.45	0.97	0.82	0.86	1.24	2.96	6.79
K_2O	0.16	1.54	1.95	2.90	0.91	2.11	2.23	2.42	2.53	0.99	0.50
Na_2O	Trace	0.58	1.04	1.47	0.20	1.98	1.36	1.04	0.53	1.04	3.45
P_2O_5	0.05	0.02	0.10	0.18	0.16	0.16	0.12	0.09	2.73	0.38	0.25
SO_3	0.05	0.07	0.08	0.12	0.07	0.21	0.12	0.11	0.78	1.66	3.14
Nitrogen	0.02	0.02	0.16	0.10	0.27	0.08	0.17	0.09	—	—	—

Note: Values are given as percentages.

From Radian Corporation, Study of Non-Hazardous Wastes From Coal-Fired Electric Utilities, EPA-68-02-2608, Washington, D.C., 1978.

Table 29
CONCENTRATIONS OF ESSENTIAL ELEMENTS IN FLY ASH AND SOIL[38,57]

	Fly ash	Common farm soil
Macronutrients, % (dry weight)		
Calcium	0.11—22.2	0.07—3.6
Magnesium	0.04—7.6	0.12—1.5
Nitrogen	0.02 (or trace)	0.02—0.5
Phosphorus	0.04—0.8	0.01—0.2
Potassium	0.15—3.5	0.17—3.3
Sulfur	0.1—1.5	0.01—0.2
Micronutrients, ppm (dry weight)		
Boron	10—618	50
Chlorine	13—460	50
Cobalt	7—520	8
Copper	14—2800	50
Iron	1—29%	2.5%
Manganese	58—3000	2500
Molybdenum	7—160	2
Zinc	10—3500	100

regions where acid soils are common, but disadvantageous in the western U.S., where soils are calcareous and already alkaline. Shifts in soil pH which often attend ash application are likely to change the availability of micronutrients.[59] Since fly ash contains boron, use of fly ash in agriculture can result in toxicity to vegetation.[50,60] Soluble salts contained in ash tend to decrease the osmotic (diffusion) potential of soil water and inhibit water uptake by plant roots. Continued inhibition of water uptake can result in reduced productivity.

Metal availability is a major concern of many scientists researching ash applications to agriculture. In general, metal toxicity in plants is implied, if observations of injury or reduced yield are correlated with elevated tissue concentrations of a particular element. Higher concentrations of aluminum, boron, selenium, strontium, sulfur, and molybdenum have been identified in tissues of plants grown on fly-ash-amended soils, compared to concentrations measured in plants on unamended soils. Studies have further identified bioaccumulation of arsenic, barium, cesium, rubidium, tungsten, and vanadium in plants grown on fly ash-amended soils. Of these twelve elements, boron has been the most common cause of reduced yields in experiments with fly ash-soil mixes.[48,61]

2. Fly Ash as a Liming Material

Liming materials are used to raise soil pH of acid soils to increase crop yields. For most crops, a soil pH between 6 and 7 is considered optimum. Investigations of the liming potential of coal ash have been limited for the most part to alkaline fly ashes.

The neutralization capacity of fly ash from different sources exhibits wide variation. Neutralizing effects of two fly ashes (Lansing and Portage) and a high-grade agricultural lime were compared (see Figure 28). The Lansing ash was approximately 40% as effective as the lime, requiring roughly two and a half times the applications rate of lime to elevate soil pH from 4.6 to about 6. The Portage ash required about three times

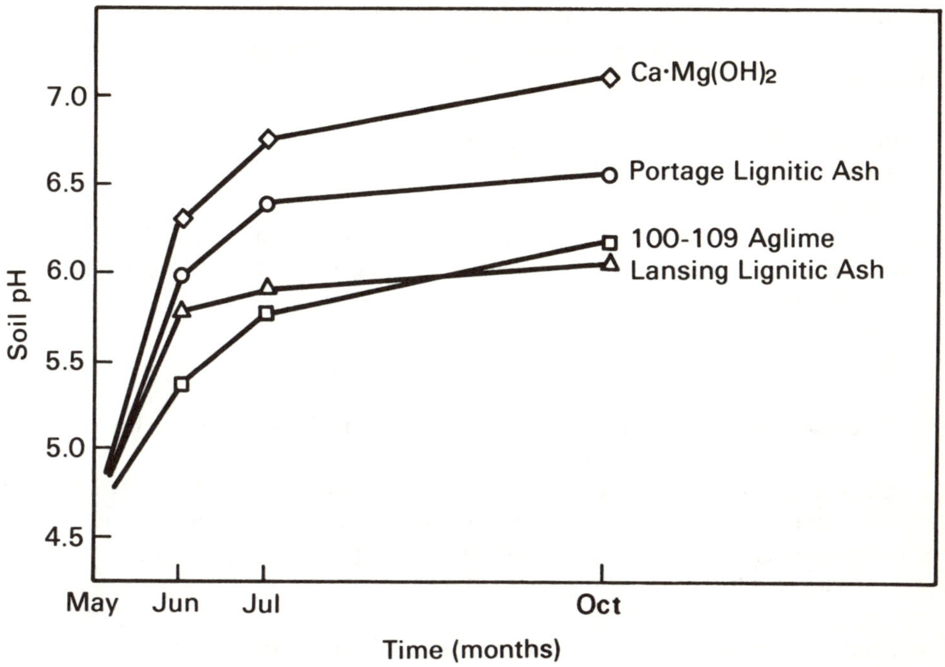

FIGURE 28. Comparison of liming materials and lignite coal fly ash as measured by pH of soils.[62]

the application rate of lime. However, both ashes had a greater initial reaction rate than the agricultural lime. An earlier reaction rate may be of potential benefit to the farmer, who could delay crop selection decisions (for reasons of market, etc.) to a later date, knowing optimum soil pH could be achieved relatively quickly following ash application.[62]

Some experimentation with flue gas desulfurization (FGD) sludge as a liming agent has been performed. The high calcium and sulfate contents of FGD sludge offer a good source for agricultural uses; however, preliminary results have been inconclusive.

Difficulties with using the coal ash in amounts required to neutralize soil acidity include metals toxicity, salinity, and reduced phosphorus availability. Significant reductions in phosphorus availability in fly-ash-amended soil have been reported,[59] limiting ash application rates.

3. Fly Ash Use in Land Reclamation

Reclamation of surface-mined lands in the eastern U.S. has made use of fly ash. Morgantown Energy Research Center in West Virginia has used fly ash to increase the pH of spoil (soil which has been damaged by mining activities) at a number of field test plots located on surface-mined lands. In most cases, surface soil pH values were increased to above 6. Subsurface soil pH values, however, remained acidic. The amounts of fly ash required to correct mine spoil acidity are large compared to limestone. Estimates range from 200 to 400 tons ash for an acre of very acid West Virginia or Pennsylvania mine spoil. In general, fly ash was found to be roughly one-tenth as effective as limestone.[63]

Reclamation of mined lands in western states involves problems associated with high sodium (Na) levels, alkalinity, and/or salinity. In portions of North Dakota, Saskatchewan, and Alberta, mine spoils are comprised of bentonitic clay which, with wetting and drying, forms a hard crust adverse to plant growth. A Canadian coal company is currently testing bottom ash as an aid in revegetating such spoils. Initial results indicate that successful growth of grains and legumes can be achieved on a 2 to 5-in. thick layer of bottom ash overlying clay spoil.

C. Structural Fill and Embankment Material

Compacted fly ash has a relatively low unit weight — an advantage when constructing a fill or embankment on top of a soft or weak subgrade. The applied weight of compacted fly ash and, therefore, resulting settlement can be substantially reduced.

Important engineering properties to consider when using fly ash in embankment and structural fill construction are

- Unit weight — Fly ash, when compacted, has a lighter unit weight than natural soils.
- Compaction characteristics — Fly ash can be easily handled and well-compacted at intermediate moisture contents.
- Shear strength — Well-compacted fly ash embankments have been shown to exhibit strengths comparable to normal soils used in earthwork operations.
- Permeability — Compacted fly ash has a low permeability, generally comparable to a clay or silty clay soil.

Examples in the construction of highway and railroad fills or embankments using fly ash are

- Melvin E. Amstutz Expressway (Waukegan, Ill.) — The project was completed in 1974 and required 246,000 yd^3 of fly ash.
- County Route 60/12 (Kanawha County, W. Va.) — Approximately 20,000 yd^3 of fly ash were used in the project which was done during the 1977 construction season.
- ASB bridge approach (North Kansas City, Mo.) — This project was completed during the 1983 construction season and involved the placement and compaction of approximately 140,000 yd^3 of ponded fly ash.

To date, fly ash has been used as structural fill or embankment material in at least eight states.[64] Two of these states, Missouri and Ohio, have prepared specifications for the use of fly ash as an embankment construction material.[64]

1. Pozzolanic Base Course Materials

One of the most successful applications of fly ash in transportation construction is in pozzolanic base course mixtures. More than 30 million tons of pozzolanic base course materials have been constructed in the U.S. over the past 30 years. The majority of these materials have involved the placement of lime-fly-ash aggregate. About 240,000 tons of fly ash were used in various road base applications in 1983. These mixtures typically consist of lime-fly-ash aggregate (LFA), lime-cement fly-ash aggregate (LCFA), or cement-fly-ash aggregate (CFA), combined with water in the proper proportions and compacted to form dense, stable bases.

LFA mixtures serve as the base or subbase material for both flexible and rigid pavement systems. A schematic diagram of a typical LFA process is provided in Figure 29. In the pavement systems, LFA is always covered with a wearing surface. In-service pavements built with LFA mixtures have base thicknesses which range from 5 to 6 in. for light-duty pavements, 6 to 10 in. for medium- to heavy-duty pavement for highways, and up to 30 in. for heavy-duty airport pavements.[65] The largest single project

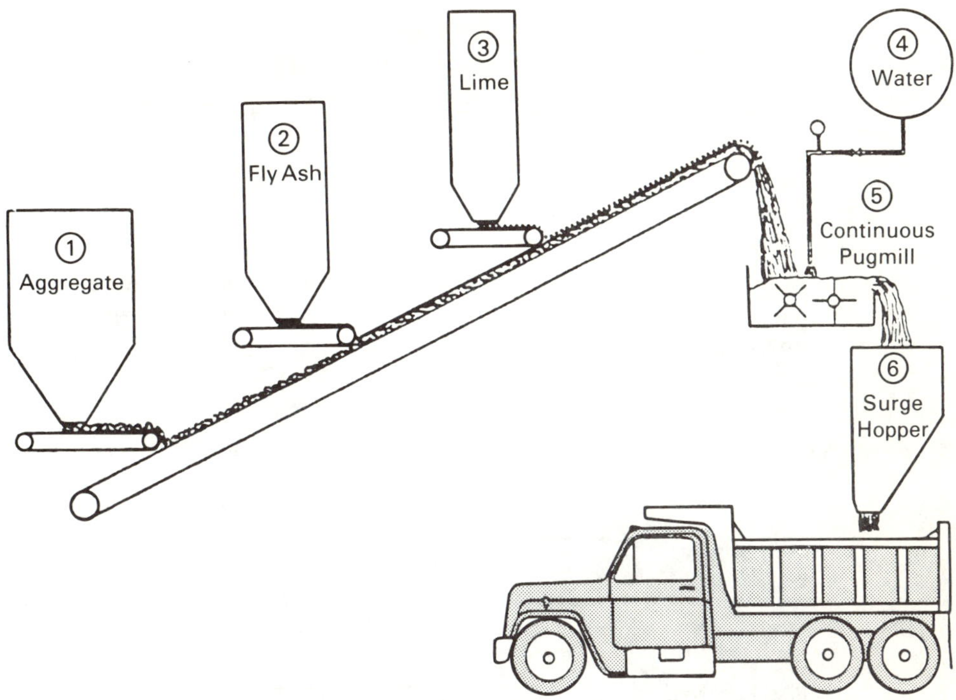

FIGURE 29. Schematic showing a lime-fly ash concrete preparation system.

involving the use of fly ash in base and subbase construction is the Newark, N.J. Airport runways, taxiways, and aprons. The redevelopment of Newark Airport called for placing approximately 2,000,000 yd^2 of new pavement.[66]

Engineers working on the project used laboratory research to develop the proper mix, which was 2.8 to 3.6% hydrated lime, 0.7 to 0.9% Portland cement, and 12 to 14% fly ash by dry weight. The remainder was hydraulic fill sand. Cross sections and mix proportions for this project are presented in Figure 30. In all, a total of 2 million tons of fly ash from Consolidated Edison's Astoria Plant were used.[66]

Lime-fly-ash aggregate (LFA) and certain other types of pozzolanic roadbase materials have been produced and used to some extent in more than a dozen states. The states in which the largest quantities have been used in highway construction for the longest period of time are Illinois, Ohio, and Pennsylvania. Other states in which the use of pozzolanic base is increasing are Missouri, North Dakota, and Texas. The material has also been used by the Federal Aviation Administration (FAA) in a number of airport paving projects, and by a number of railroad companies as a base course for the support of railroad beds.[64]

In addition to federal highway (Figure 31), state highway (Figure 32), and airfield paving (Figure 33) projects, pozzolanic base materials have been used even more extensively by municipalities (Figure 34) and private developers (Figures 35 to 36), who have found such materials to be economical and dependable alternatives to asphalt or crushed stone base.

The relative proportions of lime and/or cement, fly ash, and aggregate in pozzolanic mixtures varies, depending on the materials used and their characteristics, as well as the design criteria. Typically, LFA mixtures contain from 2 1/2 to 4% lime, and from 10 to 25% fly ash. In some cases, small quantities, i.e., 0.5 to 1.5%, of Type I Portland cement have been added during late-season construction to accelerate the initial strength gain.

Mix Designation and Symbol		Approximate Mix Proportion Ingredients (%)				
		Lime	Cement	Fly Ash	Sand	Stone
Mix A		3.6	0.9	12	53.5	30
Mix B		3.2	0.8	14	82	0
Mix C		2.8	0.7	14	82.5	0

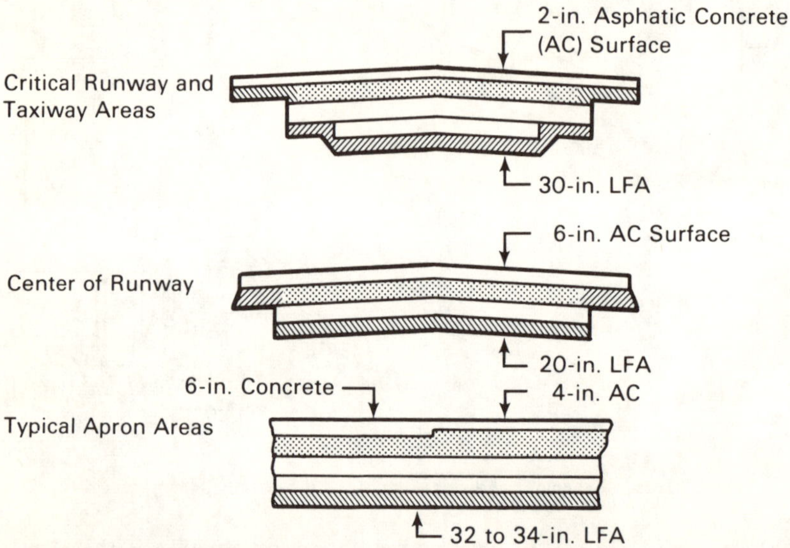

FIGURE 30. Schematic showing the layering of fly ash in airport construction.

FIGURE 31. Interstate highway in Virginia where fly ash was used in the construction.

2. Use of Boiler Slag in Pozzolanic Base

The first LFA compositions in the Chicago area were mixed in place using boiler slag as the aggregate. In 1955, the first plant-mixed LFA material also used boiler slag. These early mixtures contained on the average 5% by weight of hydrated lime, 35% fly ash, and 60% boiler slag. Cores were taken from these mixtures at various ages, and ultimate compressive strengths as high as 500 psi were measured.

FIGURE 32. Fly ash and bottom ash being used as subbase and asphalt admixture in highway construction.

FIGURE 33. An airport runway where fly ash and bottom ash were used.

FIGURE 34. Fly ash used as a basecourse for construction of a heavy-duty road.

FIGURE 35. Fly ash used as admixture to concrete for subdivision streets and curbs.

FIGURE 36. Fly ash used as subbase for a shopping mall parking lot.

During 1979, an experimental LFA test section using boiler slag was placed on Illinois Route 9 near the Coffen Power Station in Montgomery County, Ill. The test section was approximately four miles long and used a mix containing 3% lime, 27.5% fly ash, and 69.5% boiler slag. Available core sample data indicated that average compressive strengths of 1400 psi were obtained after approximately 1 year in service.

3. Use of Bottom Ash in Pozzolanic Base

In 1981, a 5-mile road demonstration project was completed near Rome, Ga., in which a 12-in. thick layer of ponded ash from the Georgia Power Company's Bowen Plant was stabilized with 8% of a byproduct lime.[64] The ponded ash was combined from two stockpiles in order to obtain a gradation similar to the Georgia Department of Transportation's specifications for a graded aggregate base. Cores taken 6 weeks after construction had average compressive strengths greater than 700 psi.

The first known large-scale application of a Portland cement-stabilized bottom ash base course in the U.S. was the relocation and reconstruction of West Virginia Route 2 south of Wheeling during the 1971 to 1972 construction seasons. The cement-treated base course for this 4-mile long project was constructed using 46% by dry weight of cyclone boiler slag from the Ohio Power Company's Kammer Plant in Captina, W.Va. and 54% bottom ash from the Appalachian Power Company's Mitchell Station in Moundsville, W.Va. The aggregate blend was stabilized with 5% Type I Portland cement, by weight of total aggregate in the mix.

Since 1974, over 300 miles of secondary roads in West Virginia have been reconstructed using cement-treated bottom ash. Most of these roads were primarily dirt and gravel subbase, with traffic counts ranging from 150 to 1,500 vehicles per day. Most of the bottom ash for these projects came from either the John Amos Plant in St. Albans, W. Va. or the Kanawha River Plant in Glasgow, W.Va. Not a single base failure was found during visual inspection of 180 miles of roadway using cement-treated bottom ash in the spring of 1978. An outstanding example of cement-stabilized bottom ash subbase is a project on Route 34 in Putnam County near Charleston, using ash from the John Amos Plant. A 6-in.-thick bottom ash subbase was placed and compacted. Successive 6-in.-thick lifts of cement-treated bottom ash base were then placed. Core specimens taken from 3 typical sections after less than 2 years in service showed average compressive strengths of 1,322 psi.[67]

D. Fly Ash as Mineral Filler in Asphalt

Over the past 40 years, the Detroit Edison Company has sold nearly 1.5 million tons of fly ash to the asphalt paving industry. Since 1969, the North Dakota Highway Department has utilized over 40,000 tons of lignite fly ash as a mineral filler. Fly ash was also used as a mineral filler in lieu of Portland cement in the placement of 35,000 tons of open-graded asphalt overlay surface on the north-south runway at the Sioux City, Iowa Municipal Airport.[64]

Since 1970, an average of 140,000 tons per year of fly ash has been used as mineral filler in the U.S. Several million tons of fly ash have been used as mineral fillers in more than 22 states over the past 40 years with satifactory results. Results of the 1983 NAA state highway survey show that at least 20 states are using fly ash as a mineral filler in asphalt paving in state highway projects. At least eight states have a specification for such use which includes fly ash.[64]

E. Soil Stabilization

Stabilization measures can improve the properties of an unsuitable soil for engineering applications by:

1. Improving inherent soil properties, such as decreasing the plasticity index or volumetric shrinkage.
2. Cementing individual soil grains to increase strength and durability characteristics.

Chemical stabilizers traditionally used for such purposes include cement, lime, lime-fly ash, bituminous materials, and chemicals such as calcium or sodium chloride. Normally, cement stabilization is used where coarse-grained soils are involved. With fine-grained soils, it is well known that the addition of lime or lime-fly ash to clays, silts, and even fine-to-medium sands, has a beneficial stabilization effect.

Examples of projects involving the use of fly ash in lime-fly-ash soil stabilization are

- Trident Base — Savannah, Ga. — During 1982, 7.7 miles of native sandy organic clay subgrade soil was stabilized with a blend of 3% Portland cement and 20% fly ash construction of base access roads.
- State Route 63 — Jackson County, Miss. — A 7.5-mile-long, lime-fly-ash soil stabilization project was completed during 1983 in Mississippi, as a demonstration program to stimulate the use of fly ash in various phases of highway construction. Fly ash was used to stabilize A-2, A-3, and A-4 soil types. Stabilization proportions consisted of one part lime to four parts fly ash.
- Beech Aircraft Plant — Wichita, Kan. — During October, 1982, more than 6,000 tons of fly ash from Kansas City Power and Light Company's Iatan Power Plant was used to stabilize the subgrade soils under the hangars, aprons, and parking lots in connection with an expansion of the Beech Aircraft Facility.[64]

F. Grouting

Grouting can be defined as the introduction of appropriate materials (whether as suspensions, solutions, or emulsions) into places of normal inaccessibility, either by pressure injection or gravitational flow, to effect improvements in structural or geological properties. The properties to be improved are strength, bearing capacity, impermeability, and resistance to chemical attack. Grouting can be accomplished in a variety of ways, including void and fissure filling and sealing, in filling of old sewers and conduits, slab jacking, and backfilling around walls and pipes.

The addition of fly ash to lime slurry grout is particularly useful where *in situ* soils are interlayered with granular materials. The lime/fly ash slurry is injected into the soil under pressure, forming a network of slurry seams through the soil. The lime/fly ash slurry seams react with adjacent soil to form strong, relatively impervious tensile membranes locked into the soil mass. Lime/fly ash slurry pressure-injection grouting has been used to stabilize soils for building foundations, railroad beds, streets, runways, and slopes and levees.

The most important uses for cement grouts are in pavement slab jacking, and in subsealing and stabilization of concrete pavements by filling voids beneath the pavement. Generally, sand/cement grouts are used for this purpose, sometimes with the addition of bentonite. The use of fly ash in cement grout slurries is a fairly recent development. Typical mix designs consist of one part Portland cement to three parts fly ash by volume, but may vary depending on fly ash characteristics.

G. Bottom Ash and Slag Aggregates in Bituminous Paving

Power plant aggregate use in bituminous paving applications has encompassed four different types of paving mixtures:

1. Bituminous stabilized cold mix base courses
2. Cold mix wearing surfaces
3. Conventional hot mix bituminous paving
4. Bituminous surface treatments

The use of bottom ash or boiler slag in each of these types of uses is described briefly below.

1. Bituminous Stabilized Cold Mix Base Courses

Some 45 to 50 miles of light-duty, rural, secondary roads in West Virginia were reconstructed during the summer of 1972, using bituminous-stabilized power plant aggregates. Both bottom ash and boiler slag were used in these secondary road projects. These base materials were cold mixed and placed directly on existing gravel or badly deteriorated chip seal surfaces in single lifts varying from 2 to 6 in. in thickness. The base materials did not receive a surface treatment until the following constructive season.[67]

Bottom ash for these mixes was obtained from the Fort Martin Station and boiler slag was obtained from the Kammer Power Plant. The dry bottom ash was used without blending with other aggregate, since it was able to satisfy gradation specifications. The bottom ash was blended with asphalt emulsion in pugmill with a design asphalt content of 7%.

2. Cold Mix Wearing Surfaces

The most extensive use of bottom ash in bituminous paving has been in West Virginia where, since 1972, bottom ash has been cold mixed with 6 to 7% by weight of emulsified asphalt and used in the paving of secondary roads (Figure 37). In some cases, the ash is also blended with boiler slag. More than 200 miles of low-volume traffic roads in the northern part of the state have been improved with these cold mix compositions, which are referred to as "asphalt". Similar applications have also been made in eastern Ohio.[64]

3. Conventional Hot-Mix Bituminous Paving

There are numerous examples of the use of power plant aggregates in hot-mix bituminous paving projects. Most of these involve the use of boiler slag. Boiler slag has been used to a much greater extent in bituminous paving than bottom ash. Boiler slag has been used frequently in wearing surface mixtures because of the hardness, its affin-

FIGURE 37. Road built with bottom ash.

ity for asphalt, and its dust-free surface, which aids in asphalt adhesion and resists stripping. Another of the properties of boiler slag which enhances its value as an aggregate in bituminous paving is its permanent black color, which is not affected by sun or weather. The black color also aids in the melting of snow from the road surface during the winter.

H. Ice Control Material

Bottom ash and boiler slag have both proved to be useful when spread on ice and snow-covered roads to control skidding (Figure 38). Although this is a seasonal use, it is an important reuse of coal ash.

Among the engineering properties required of an ice-control material, the following are considered most important:

- Particle size distribution — A well-graded material is preferred for ice control. Normally, a simple screening operation is all that is required for bottom ash or boiler slag to become a specification material.
- Abrasion resistance — Individual particles should not be brittle or too easily crushed under the action of moving wheels.
- Unit weight or density — Ice control material should be of low to moderate unit weight, so that, when particles are propelled into the windshields of other vehicles, broken glass will not result.
- Absence of corrosive substances — Power plant aggregates are essentially inert materials. Consequently, exposure to power plant aggregates is not detrimental to auto bodies in the same way as road salts are.

There is a substantial demand for ice-control materials in most northern and midwestern states. During the past few winters, one utility company in Pennsylvania has sold as much as 50,000 tons per year of screened bottom ash for use as an ice-control material. The Pennsylvania Department of Transportation (Penn DOT) has developed a specification to prequalify sources of the material for annual bid.

I. Lightweight Aggregate Manufacturing

There are three basic categories of lightweight aggregates in the U.S.[68]:

FIGURE 38. Boiler slag and bottom ash used for ice-control on highways in the Northeast U.S.

1. Natural — Aggregate from naturally occurring materials such as pumice, scoria, volcanic cinders, and tuff.
2. Byproducts — Aggregates generally derived from the combustion of coal or coke including expanded slag, cinders, and sinter fly ash.
3. Ultra-lightweights — Natural mineral ores like perlite and vermiculite processed into aggregate materials.

The primary uses for lightweight aggregates are in blocks and structural and precast concrete where minimum weight (lb/ft^3) is important. When used in these applications, typical standards for production and usage are ASTM C330-77, "Lightweight Aggregates for Structural Concrete," and ASTM C331-77, "Lightweight Aggregates for Concrete Masonry Units".

The production of lightweight aggregate from fly ash is basically a two-stage process in which pellets are formed and then sintered.

First, the fly ash, water, finely divided fuel, and in some cases additives, are combined in a rotating pan or drum and mixed. The mixture is then put into a grate, and the pellets are ignited and burned at about 1000 to 1300°C. The resulting aggregates can then be crushed to the desired size.[69,70]

Its product, Lytag, has been used throughout Great Britain. Up until February 1978, fly ash lightweight aggregates were produced in the U.S. by Penn Virginia Materials.[70] Due to changes in ash handling systems resulting in changes in fly ash characteristics, and difficulties in obtaining a dependable supply of ash, Penn Virginia ceased production. Another company which produced fly ash lightweight aggregate is the Waylite Company of River Rouge, Mich. In 9 years, this company sintered over a million tons of fly ash. Structures using this lightweight concrete include: the 24-story Walker Asler Building in Detroit, and a 7-building complex comprised of 6- to 7-story apartment houses in Springfield, Michigan.[71]

J. Mineral Resources

Minerals found in fly ash originate as part of the inorganic matter from silts and clays deposited in the ancient sedimentary environments, and by metamorphic and exchange processes occurring in the coal seam. Most of the metallic elements in coal remain as metal oxides in ash after combustion.

Fly ash is considered to be more suitable for extraction processes than bottom ash. Bottom ash tends to be nonuniform in particle size and may include fines, cinders, clinkers, and slags. Fly ash is better sorted for size and can be more simply metered for processing. Fly ash particle size is small, so that a large amount of surface is available for contact in chemical extraction processes. For these reasons, extraction process development has been oriented toward fly ash.

The major metallic constituents of fly ash, in order of decreasing concentration are: silicon, aluminum, iron, calcium, magnesium, and titanium. Oxides and silicates of these metals typically exceed 90% of fly ash by weight. Aluminum, iron, and titanium metals are of direct industrial importance, while calcium silicate is used in the manufacture of Portland cement.

1. Magnetic Extraction

Most of the iron in fly ash is present in the form of magnetite, which is highly magnetic. This iron oxide is directly extractable by electromagnets. Rates of recovery are high, and an estimated 10 to 15% by weight of bituminous fly ash can be removed by this process.[72] The magnetic removal of all iron from fly ash is inhibited by the occurrence of some of the iron within larger nonmagnetic particles. The magnetic fly ash fraction is usable as an iron ore but has greater value as a heavy medium material for coal beneficiation. Tests at the Ames Laboratory of the USDOE show that the densities of suspensions of magnetic fly ash fractions are equivalent to those of mined suspensions (1.3 to 1.4 g/cm^3). Grinding and washing the magnetic fly ash fraction produces a better product because the smaller particles settle from suspension at a slower rate.

At least one company, Penn Virginia Materials Corporation, has produced heavy medium material from fly ash, but production has ceased. The G. and W. H. Corson Company developed a Roto-Flux® magnetic separator which was a model for the laboratory scale separator used at the Ames Laboratory.[73]

In addition to reducing the volume of ash, magnetic extraction of iron also concentrates other minerals remaining in fly ash. Because iron interferes in most other extraction processes, its removal facilitates extraction of other minerals from the remaining fly ash.

2. Alumina

The recovery of alumina (Al_2O_3) from fly ash is a theoretical use receiving wide attention. The aluminum content of most fly ashes (16.2 to 29.6%) is similar. Microscopic examination of fly ash reveals that it is composed of glassy spheres. Because glasses are resistant to chemical alteration, strong chemical conditions are required to leach metals from fly ash at industrially usable levels. Processes such as the Pechiney process, which use a very strong, high-temperature, sulfuric acid solution to extract aluminum sulfate from clays and crushed shales, are not as suitable for extracting alumina from glasses. A study conducted at Oak Ridge National Laboratory[74] (ORNL) found maximum recoveries occurred in concentrated acid. Recoveries were 56% for aluminum, 73% for iron, and 40% for titanium.[74]

The Bayer process, used in the refining of alumina from bauxite, does not appear to be suitable for application to coal ashes. In the Bayer process, precipitation from a solution of aluminate and silicate does not produce a sufficiently pure aluminum hydroxide end-product when a high silica feed such as fly ash is used. However, several metal extraction processes suitable for use with fly ash have been developed and tested at bench or pilot scale levels.[74,75]

The sintering processes consist of heating pellets of a mixture of fly ash, soda ash, and lime or limestone to a temperature of from 700° to 1400°c. The pelletizing creates close physical contact for the reaction which produces insoluble silicates and soluble aluminates. Leaching followed by precipitation then produces a hydrated alumina. Several variations of this process are

- Lime-sinter — A process used for extracting alumina from bauxites high in silica, the Pedderson or lime-sinter process, has been applied to fly ash. The alumina is released by supplying a reagent (lime), which has a greater affinity for silica than does alumina.
- Lime-soda-sinter — Developed in Hungary as the Tatabany process, this process adds sodium carbonate along with limestone to fly ash prior to sintering. The formation of soluble sodium aluminates in addition to calcium aluminates raises the efficiency of this process. Canon et al. realized alumina recoveries of 60 to 70%.[74]

  Burnet and Murtha at the Ames Laboratory obtained recoveries of alumina varying from 74 to nearly 100% depending upon conditions such as the particle size of the sintered material.[75] The Ames process is carried out at about 1200°C and produces a clinker which is ground before extraction. A 3% solution of sodium carbonate is used to leach sodium aluminate from the ground clinker.
- Calsinter — At ORNL, sintering of a mixture of fly ash, gypsum, and limestone at 1000 to 1200°C has been conducted for metal extraction.[74] This process recovers less than 65% of the aluminum in fly ash at 800°C, but recovers up to 97% of the aluminum after a 1300°C sintering. Leaching of the ground clinker with sulfuric acid produces a solution of aluminum sulfate (alum). Alumina may be obtained by crystallization and calcining of the aluminum sulfate.
- High-temperature chlorination — At the USDOE Ames Laboratory, workers are in the early stages of development of a high-temperature chlorination extraction process called "HiChlor".[75] This process is based on the formation of volatile metal halides at high temperatures, and the subsequent separation of these gases. In the process, fly ash reacts with chlorine gas in the presence of a reducing agent such as carbon to form gaseous metal chlorides.

K. Artificial Reefs

A novel use of coal fly ash was demonstrated in a study conducted for EPRI where blocks from coal-ash mixtures were constructed and laid on the ocean floor off Long Island, N.Y.[76] The ash blocks formed a constructed reef on the ocean floor and resulted in providing a habitat for the aquatic organisms. Additional attempts are being made in Wisconsin, where an artificial reef made of coal wastes will be constructed in Lake Michigan. Similarly in Florida, an artificial reef in the ocean is being developed.

VIII. REUSE OF FGD SLUDGES

The FGD sludges are relatively new solid wastes produced in the electric utility industry. These wastes are composed of large fractions of $CaSO_4$ and $CaSO_3$. The most likely uses of the FGD wastes are for the production of commercial grade gypsum which can then be used in the manufacture of wall boards. In addition, limited testing has been done to produce elemental sulfur, sulfur dioxide, or sulfuric acid from a recovery plant attached to an FGD system. At the present time, FGD sludges are not used for gypsum, elemental sulfur, or sulfuric acid production. It is expected that, in the future, recovery of sulfur compounds from FGD wastes will be practiced if the economics and market demands become favorable.

APPENDIX A
List of Coal-Fired Power Plants Operating in the U.S., 1984

State	Plant name	No. of units	Year of first operation	Coal type	Megawatts capacity
AL	Colbert	5	54,55,55,55,62	Bituminous	1396
AL	Green County	2	65,66	Bituminous	500
AL	Widows Creek	8	52,52,52,53,54,54,60,64	Bituminous	1975
AL	Miller	1	78	Bituminous	660
AL	Barry	5	54,59,69,71	Bituminous	1525
AL	E.C. Gaston	5	60,60,61,62,74	Bituminous	1880
AL	Gorgas Two	6	44,51,52,56,58,72	Bituminous	1281
AL	Tombigbee	3	69,79,80	Bituminous	585
AR	Flint Creek	1	78	Subbituminous	558
AR	White Bluff	2	80,81	Subbituminous	1600
AR	Independence	1	83	Subbituminous	900
AZ	Coronado	2	79,80	Subbituminous	822
AZ	Apache	2	79,79	Bituminous	398
AZ	Navajo	3	74,75,76	Bituminous	2409
AZ	Cholla	4	62,78,80,81	Bituminous	1017
CO	Cherokee	4	57,59,62,68	Bituminous	710
CO	Valmont	1	64	Subbituminous	166
CO	Arapahoe	4	50,51,51,55	Bituminous	232
CO	Craig	2	80,79	Subbituminous	892
CO	Pawnee	1	81	Subbituminous	552
CO	Haydan	2	65,76	Bituminous	465
CO	Comanche	2	73,75	Subbituminous	765
CO	Drake	3	62,68,74	Subbituminous	294
CO	RD Nixon	1	80	Bituminous	230
DE	Indian River	4	57,59,70,80	Bituminous	782
FL	Deerhaven	1	81	Bituminous	250
FL	Lansing Smith	2	65,67	Bituminous	305
FL	Crystal River	3	66,69,77	Bituminous	1853
FL	Crist	4	59,61,70,73	Bituminous	919
FL	F.J. Gannon	2	65,67	Bituminous	653
FL	Big Bend	3	70,73,76	Bituminous	1336
FL	McIntosh	1	82	Bituminous	364
FL	Seminole	1	83	Bituminous	680
GA	Bowen	4	71,72,74,75	Bituminous	3160
GA	Port Wentworth	3	58,61,65	Bituminous	207
GA	Jack McDonough	2	63,64	Bituminous	490
GA	Yates	7	50,50,53,57,58,74,74	Bituminous	1250
GA	Hammond	4	54,54,55,70	Bituminous	800
GA	Mitchell	3	48,49,64	Bituminous	170
GA	Scherer	1	82	Bituminous	818
GA	Harllee Branch	4	65,67,68,69	Bituminous	1540
GA	Wansley	2	76,78	Bituminous	1730
IL	Lakeside	4	49,53,60,65	Bituminous	116
IL	Fisk	1	59	Bituminous	374
IL	Crawford	2	58,61	Bituminous	597
IL	Kincaid	2	67,68	Bituminous	1320
IL	Duck Creek	1	76	Bituminous	416
IL	Grand Tower	2	51,58	Bituminous	200
IL	Newton	2	77,82	Bituminous	1235
IL	Waukegan	3	52,58,62	Bituminous	802
IL	Wood River	2	54,64	Bituminous	500
IL	Havana	1	78	Bituminous	450
IL	Joppa	6	53,53,54,54,55,55	Bituminous	1098
IL	Hutsonville	2	53,54	Bituminous	150
IL	Coffeen	2	65,72	Bituminous	1005

APPENDIX A (continued)
List of Coal-Fired Power Plants Operating in the U.S., 1984

State	Plant name	No. of units	Year of first operation	Coal type	Megawatts capacity
IL	Meredosia	3	48,49,60	Bituminous	354
IL	Ed Edwards	3	60,68,72	Subbituminous/ bituminous	700
IL	Hennepin	2	53,59	Bituminous	306
IL	Baldwin	3	70,73,75	Bituminous	1892
IL	Dallman	3	67,72,78	Bituminous	352
IL	R.S. Wallace	3	49,52,58	Bituminous	240
IL	Powerton	2	72,75	Bituminous	1785
IL	Vermillion	2	55,56	Bituminous	182
IL	Joliet	3	59,65,66	Bituminous	1680
IL	Will County	4	55,55,57,63	Bituminous	1269
IL	Marion	4	63,63,63,78	Bituminous	272
IN	Tanners Creek	4	51,52,54,64	Bituminous	1100
IN	R.H. Gallagher	4	58,59,60,61	Bituminous	600
IN	Gibson	5	75,76,78,79,82	Bituminous	3340
IN	R.M. Schahfer	3	77,79,83	Bituminous	1425
IN	Clifty Creek	6	55,55,55,55,55,56	Bituminous	1303
IN	Stateline	2	55,62	Bituminous	614
IN	Dean H. Mitchell	3	56,59,59	Bituminous	414
IN	Michigan City	3	50,51,74	Bituminous	661
IN	E.W. Stout	3	58,61,73	Bituminous	698
IN	H.T. Pritchard	4	51,53,53,56	Bituminous	281
IN	Ratts	2	68,68	Bituminous	233
IN	Petersburg	3	67,69,77	Bituminous	1255
IN	Bailly	2	68,68	Bituminous	615
IN	A.B. Brown	1	79	Bituminous	265
IN	Merom	2	82,83	Bituminous	980
IN	Breed	1	60	Bituminous	496
IN	Cayuga	2	70,72	Bituminous	1062
IN	Wabash River	6	53,53,54,55,56,68	Bituminous	962
IN	Warrick	1	70	Bituminous	323
IN	Culley	3	55,66,73	Bituminous	415
IN	Edward Sport	3	49,49,51	Bituminous	108
IA	Muscatine	5	49,49,51,?,?	Bituminous	274
IA	Lansing	4	48,49,57,77	Subbituminous	368
IA	ML Kapp	2	47,67	Bituminous	237
IA	Burlington	1	68	Bituminous	212
IA	Prairie Creek	4	50,50,57,67	Bituminous	244
IA	Des Moines	2	54,64	Bituminous	189
IA	Council Bluffs	3	54,58,78	Subbituminous	805
IA	Riverside	3	37,49,61	Bituminous	201
IA	Ottumwa	1	81	Subbituminous	807
IA	George Neal	4	64,72,76,79	Subbituminous	1584
IA	Ames	4	50,58,68,82	Subbituminous/ bituminous	124
IA	Sutherland	4	50,55,55,61	Bituminous	156
KY	East Bend	1	81	Bituminous	648
KY	Ghent	3	74,77,81	Bituminous	1533
KY	Elmer Smith	2	64,74	Bituminous	416
KY	Green	2	80,81	Bituminous	480
KY	Reid	1	65	Bituminous	80
KY	Mill Creek	4	72,74,78,82	Bituminous	1703
KY	Cane Run	6	54,56,58,62,66,69	Bituminous	1017
KY	Big Sandy	2	63,69	Bituminous	1097
KY	H.L. Spurlock	2	77,81	Bituminous	813

APPENDIX A (continued)
List of Coal-Fired Power Plants Operating in the U.S., 1984

State	Plant name	No. of units	Year of first operation	Coal type	Megawatts capacity
KY	Shawnee	10	53,53,53,54,54,54,54,55,55,56	Bituminous	1750
KY	EW Brown	3	57,63,71	Bituminous	694
KY	Green river	4	50,50,54,59	Bituminous	263
KY	Paradise	3	63,63,70	Bituminous	2558
KY	J.S. Cooper	2	64,69	Bituminous	321
KY	Coleman	3	69,70,72	Not known	519
KY	Dale	4	54,54,57,60	Bituminous	172
KY	Henderson	4	73,73,57,68	Bituminous	341
KY	Paddys Run	1	52	Bituminous	75
KS	Riverton	2	50,54	Bituminous	88
KS	Kaw	3	55,57,62	Bituminous	235
KS	Nearman Creek	1	81	Subbituminous	235
KS	Quindaro	2	66,71	Bituminous	239
KS	La Cygne	2	73,77	Subbituminous	1510
KS	Jeffrey	3	78,80,83	Subbituminous	2160
KS	Lawrence	2	60,71	Subbituminous	517
KS	Tecumseh	2	62,57	Bituminous	232
KS	Holcomb	1	84	Subbituminous	348
LA	Big Cajun Two	3	81,81,84	Subbituminous	1695
LA	RS Nelson	1	82	Subbituminous	540
MD	CP Crane	2	61,63	Bituminous	400
MD	HA Wagner	1	66	Bituminous	359
MD	RP Smith	2	47,57	Bituminous	109
MD	Chalk Point	2	64,65	Bituminous	728
MD	Dickerson	3	59,60,62	Bituminous	588
MD	Morgantown	2	70,71	Bituminous	1148
MA	Mount Tom	1	60	Bituminous	136
MA	Brayton Point	4	63,64,69,74	Bituminous	1500
MA	Salem Harbor	3	51,52,58	Bituminous	330
MI	B.C. Cobb	5	48,48,50,56,57	Bituminous	510
MI	D.E. Karn	2	59,61	Bituminous	530
MI	J.C. Weadock	2	55,58	Bituminous	312
MI	J.H. Campbell	3	62,67,80	Bituminous	1428
MI	Belle River (?)	2	84,85	Subbituminous	1398
MI	Conners Creek	2	51,51	Bituminous	263
MI	Harbor Beach	1	68	Bituminous	121
MI	Marysville	3	42,43,47	Bituminous	200
MI	Monroe	4	71,73,73,74	Bituminous	3280
MI	River Rouge	2	57,58	Bituminous	581
MI	St. Clair	6	53,53,54,54,61,69	Bituminous	1527
MI	Trenton Channel	3	49,50,68	Bituminous	775
MI	J.B. Sims	3	61,61,83	Bituminous	101
MI	Eckert	6	54,58,61,64,68,70	Bituminous	286
MI	Erickson	1	73	Bituminous	165
MI	Presque Isle	9	55,62,64,66,74,75,78,78,79	Bituminous	628
MN	Clay Boswell	4	58,60,73,80	Subbituminous	1068
MN	Syl Laskin	2	53,53	Subbituminous	110
MN	Alan S. King	1	68	Bituminous/ subbituminous	598
MN	Black Dog	4	52,54,55,60	Bituminous/ subbituminous	445
MN	High Bridge	4	42,44,56,59	Bituminous/ subbituminous	396
MN	Riverside	4	38,31,49,64	Bituminous/ subbituminous	278

APPENDIX A (continued)
List of Coal-fired Power Plants Operating in the U.S., 1984

State	Plant name	No. of units	Year of first operation	Coal type	Megawatts capacity
MN	Sherburne County	2	76,77	Subbituminous	1440
MN	Hoot Lake	3	48,59,64	Lignite	137
MN	Silver Lake	4	48,53,62,69	Bituminous	99
MI	Jack Watson	2	68,73	Bituminous	877
MI	VJ Daniel	2	77,81	Bituminous	1096
MI	Morrow	2	78,78	Subbituminous	406
MO	New Madrid	2	72,77	Bituminous	1200
MO	Thomas Hill	3	66,69,82	Bituminous	1213
MO	Asbury	1	70	Bituminous	213
MO	Blue Valley	3	58,58,65	Bituminous	115
MO	Hawthorn	5	51,51,53,55,69	Bituminous	908
MO	Iatan	1	80	Subbituminous	674
MO	Montrose	3	58,60,64	Subbituminous	564
MO	Sibley	3	60,62,69	Bituminous	523
MO	Sikeston	1	81	Bituminous	235
MO	James River	5	57,57,60,64,70	Bituminous	253
MO	South West	1	76	Bituminous	194
MO	Lake Road	2	51,67	Bituminous	113
MO	Labadie	4	70,71,72,73	Bituminous	2482
MO	Mermec	4	53,54,59,61	Bituminous	923
MO	Rush Island	2	76,77	Bituminous	1241
MO	Sioux	2	67,68	Bituminous/ subbituminous	1100
MT	Coalstrip	3	75,76,84	Subbituminous	1494
MT	J.E. Corette	1	68	Subbituminous	173
NE	L.D. Wright	3	58,63,76	Bituminous	136
NE	Platte	1	82	Not known	147
NE	Hastings	1	81	Subbituminous	76
NE	Gerald Gentleman	2	79,81	Subbituminous	1362
NE	Kramer	3	49,49,51	Subbituminous	121
NE	Sheldon	2	61,66	Bituminous	228
NE	Nebraska City	1	79	Subbituminous	565
NE	North Omaha	5	54,59,59,63,68	Bituminous	638
NV	Reid Gardner	4	65,68,76,83	Bituminous/ subbituminous	616
NV	North Valmy	1	81	Bituminous	254
NV	Mohave	2	70,71	Subbituminous	1636
NH	Merrimack	2	60,68	Bituminous	459
NJ	B.L. England	2	62,64	Bituminous	299
NJ	Hudson	1	68	Bituminous	660
NJ	Mercer	2	60,61	Bituminous	653
NM	Four Corners	5	63,63,64,69,70	Subbituminous	2173
NM	Plains	1	84	Subbituminous	233
NM	San Juan	4	73,76,79,82	Subbituminous	1708
NY	Goudey	2	43,51	Bituminous	103
NY	Greenidge	4	38,43,50,53	Bituminous	199
NY	Hickling	2	48,52	Bituminous	70
NY	Milliken	2	55,58	Bituminous	270
NY	Somerset	1	84	Bituminous	625
NY	C.R. Huntley	6	42,48,53,54,57,58	Bituminous	836
NY	Dunkirk	4	50,50,59,60	Bituminous	628
NY	Beebee	1	59	Bituminous	81
NY	Russell	4	49,51,53,57	Bituminous	252
NC	Asheville	2	64,71	Bituminous	413
NC	Cape Fear	2	56,58	Bituminous	328

APPENDIX A (continued)
List of Coal-Fired Power Plants Operating in the U.S., 1984

State	Plant name	No. of units	Year of first operation	Coal type	Megawatts capacity
NC	Lee	3	51,52,62	Bituminous	402
NC	Mayo	1	84	Bituminous	720
NC	Roxboro	4	66,68,73,80	Bituminous	2558
NC	Sutton	3	54,55,72	Bituminous	671
NC	Weatherspoon	3	49,50,52	Bituminous	165
NC	Allen	5	57,57,59,60,61	Bituminous	1155
NC	Belews Creek	2	74,75	Bituminous	2160
NC	Buck	4	41,42,53,53	Bituminous	370
NC	Cliffside	5	40,40,48,48,72	Bituminous	781
NC	Dan River	3	49,50,55	Bituminous	290
NC	Marshall	4	65,66,69,70	Bituminous	2000
NC	Riverbend	4	52,52,54,54	Bituminous	466
ND	Antelope Valley	1	83	Lignite	455
ND	Leland Olds	2	66,75	Lignite	656
ND	Coal Creek	2	79,81	Lignite	1100
ND	M.R. Young	2	70,77	Lignite	705
ND	Coyote	1	81	Lignite	456
ND	Haskett	2	54,63	Lignite	100
ND	Stanton	2	66,82	Lignite	187
OH	Cardinal	3	67,68,77	Bituminous	1880
OH	Miami Fort	4	49,60,75,78	Bituminous	1292
OH	W.C. Beckjord	6	52,53,54,58,62,69	Bituminous	1164
OH	Ashtabula	5	48,48,48,48,58	Bituminous	440
OH	Avon Lake	4	49,49,59,70	Bituminous	1085
OH	East Lake	5	53,53,54,56,72	Bituminous	1257
OH	Conesville	6	57,59,62,73,76,78	Bituminous	1780
OH	Picway	1	55	Bituminous	90
OH	Poston	4	49,50,52,54	Bituminous	204
OH	Hutchings	6	48,49,50,51,52,53	Bituminous	414
OH	J.M. Stuart	4	70,71,72,74	Bituminous	2440
OH	Hamilton	2	66,75	Bituminous	78
OH	Edgewater	3	24,49,57	Bituminous	211
OH	Gorge	2	43,48	Bituminous	87
OH	Niles	2	54,54	Bituminous	250
OH	R.E. Burger	5	44,47,50,55,55	Bituminous	545
OH	Toronto	3	40,49,49	Bituminous	175
OH	W.H. Sammis	7	59,60,61,62,67,69,71	Bituminous	2303
OH	General JM Gavin	2	74,75	Bituminous/ subbituminous	2600
OH	Muskingum River	5	53,54,57,58,68	Bituminous	1529
OH	Kyger Creek	5	55,55,55,55,55	Bituminous	1086
OH	Vine Street	3	62,70,71	Bituminous	72
OH	Acme	3	41,49,51	Bituminous	235
OH	Bay Shore	4	55,59,63,68	Bituminous	632
OK	Muskogee	2	78,79	Subbituminous	1030
OK	Sooner	2	79,80	Subbituminous	1030
OK	Northeastern	2	79,80	Subbituminous	945
OK	Hugo	1	82	Subbituminous	400
OR	Boardman	1	80	Subbituminous	561
PA	Cheswick	1	70	Bituminous	565
PA	Elrama	4	52,53,54,60	Bituminous	520
PA	F.R. Phillips	4	43,49,50,56	Bituminous	410
PA	Portland	2	58,62	Bituminous	404
PA	Titus	3	51,51,53	Bituminous	225
PA	Conemaugh	2	70,71	Not known	1872

APPENDIX A (continued)
List of Coal-Fired Power Plants Operating in the U.S., 1984

State	Plant name	No. of units	Year of first operation	Coal type	Megawatts capacity
PA	Front Street	5	28,44,52,53,56	Not known	120
PA	Homer City	3	69,69,77	Not known	1970
PA	Keystone	2	67,68	Not known	1872
PA	Seward	2	50,57	Not known	193
PA	Shawville	4	54,54,59,60	Not known	640
PA	Warren	2	48,49	Not known	85
PA	Brunner Island	3	61,65,69	Bituminous	1558
PA	Holtwood	1	54	Anthracite	75
PA	Martins Creek	4	54,56,75,77	Bituminous	2012
PA	Montour	2	71,73	Bituminous	1624
PA	Sunbury	4	49,49,51,53	Bituminous	410
PA	Bruce Mansfield	3	76,77,80	Bituminous	2505
PA	New Castle	5	39,47,52,58,64	Bituminous	426
PA	Cromby	1	54	Bituminous	187
PA	Eddystone	2	59,60	Bituminous	707
PA	Armstrong	2	58,59	Bituminous	361
PA	Hatsfields Ferry	3	69,70,71	Bituminous	1728
SC	H.B. Robinson	1	60	Bituminous	207
SC	Lee	3	51,51,58	Bituminous	345
SC	Canadys	3	62,64,67	Not known	489
SC	Mcmeekin	2	58,58	Not known	294
SC	Urquhart	3	53,54,55	Not known	250
SC	Wateree	2	70,71	Bituminous	772
SC	Cross	1	84	Bituminous	530
SC	Grainger	2	66,66	Bituminous	163
SC	Jeffries	2	69,69	Bituminous	345
SC	Winyah	4	75,77,80,81	Bituminous	1260
SD	Big Stone	1	75	Lignite	455
TN	Bull run	1	66	Bituminous	950
TN	Cumberland	2	73,73	Bituminous	2600
TN	Gallatin	4	56,57,59,59	Bituminous	1255
TN	John Sevier	4	55,55,56,57	Bituminous	847
TN	Johnsonville	10	51,51,52,52,52,53,58,59,59,59	Bituminous	1485
TN	Kingston	9	54,54,54,54,55,55,55,55,55	Bituminous	1723
TN	T.H. Allen	3	58,59,59	Bituminous	990
TX	Coleto Creek	1	80	Subbituminous	610
TX	W.A. Parish	4	77,78,80,82	Subbituminous	2697
TX	Fayette	2	79,80	Subbituminous	1137
TX	J.T. Deely	2	77,78	Subbituminous	836
TX	San Miguel	1	82	Lignite	447
TX	Walker County	1	80	Lignite	720
TX	Welsh	3	77,80,82	Subbituminous	1674
TX	Harrington	3	76,78,80	Subbituminous	1080
Tx	Tolk	1	82	Subbituminous	568
TX	Gibbons Creek	1	84	Lignite	443
TX	Sandow	1	81	Lignite	591
TX	Big Brown	2	71,72	Lignite	1186
TX	Martin Lake	3	77,78,79	Lignite	2379
TX	Monticello	3	74,75,78	Lignite	1979
UT	Carbon	2	54,57	Bituminous	188
UT	Gadsby	2	52,55	Bituminous	183
UT	Hunter	3	78,80,83	Bituminous	1322
UT	Huntington	2	77,74	Bituminous	893
VA	Clinch River	3	58,58,61	Bituminous	712
VA	Glen Lyn	2	44,57	Bituminous	356

APPENDIX A (continued)
List of Coal-Fired Power Plants Operating in the U.S., 1984

State	Plant name	No. of units	Year of first operation	Coal type	Megawatts capacity
VA	Potomic River	5	49,50,54,56,57	Bituminous	508
VA	Bremo Bluff	2	50,58	Bituminous	254
VA	Chesterfield	3	60,64,69	Bituminous	1240
VA	Portsmouth	1	62	Bituminous	239
VA	Possum Point	2	55,62	Bituminous	353
WA	Centralia	2	71,72	Subbituminous	1330
WV	Amos	3	71,72,73	Bituminous	2932
WV	Kanawha River	2	53,53	Bituminous	439
WV	Mountaineer	1	80	Bituminous	1300
WV	Philip Sporn	5	50,50,51,52,60	Bituminous	1105
WV	Albright	3	52,52,54	Bituminous	289
WV	Fort Martin	2	67,68	Bituminous	1152
WV	Harrison	3	72,73,74	Bituminous	2052
WV	Pleasants	2	79,80	Bituminous	1368
WV	Rivesville	2	44,51	Bituminous	110
WV	Willow Island	2	49,60	Bituminous	215
WV	Kammer	3	58,58,59	Bituminous	712
WV	Michell	2	71,71	Bituminous	1632
WV	Mount Storm	3	65,66,73	Bituminous	1662
WI	Alma	5	47,47,50,50,59	Bituminous/ subbituminous	191
WI	Genda	1	69	Bituminous/ subbituminous	345
WI	Madgett	1	79	Subbituminous	367
WI	Blount Street	4	53,48,57,61	Bituminous	158
WI	Oak Creek	8	53,54,55,57,59,61,65,67	Bituminous	1670
WI	Pleasant Prairie	2	80,84	Subbituminous	1234
WI	Port Washington	5	35,43,48,49,50	Bituminous	400
WI	Valley	2	68,69	Bituminous	280
WI	Columbia	2	75,78	Subbituminous	1024
WI	Edgewater	4	31,31,51,69	Bituminous	450
WI	Nelson Dewey	2	60,62	Bituminous/ subbituminous	227
WI	Rock River	2	54,55	Bituminous	160
WI	J.P. Pulliam	6	43,47,49,51,58,64	Bituminous	372
WI	Weston	3	54,60,81	Bituminous	456
WY	Laramie River	3	80,81,82	Subbituminous	1800
WY	Dave Johnston	4	59,61,64,72	Subbituminous	788
WY	Jim Bridger	4	74,75,76,79	Subbituminous	2034
WY	Wyodak	1	78	Subbituminous	332
WY	Naughton	3	63,68,71	Subbituminous	707

APPENDIX B
List of Plants and Type of FGD System Operations in 1984[4]

State	Plant name	Capacity [MW(e)]	FGD system type
Alabama	Tombigbee 2—3	358	Wet limestone
Alabama	Widows Creek 7—8	1125	Wet limestone
Arizona	Apache 2—3	196	Wet limestone
Arizona	Cholla 1—4	1017	Wet limestone
Arizona	Coronado 1—2	822	Wet limestone
Colorado	Craig 1—3	1340	Wet limestone

APPENDIX B (continued)
List of Plants and Type of FGD System Operations in 1984[4]

State	Plant name	Capacity [MW(e)]	FGD system type
Colorado	Rawhide 1	250	Dry lime
Delaware	Delaware City 1—3	180	Wellman Lord
Florida	Big Bend 4	417	Wet limestone
Florida	McIntosh 3	364	Wet limestone
Florida	Seminole 1	680	Wet limestone
Iowa	Muscatine 9	166	Wet limestone
Illinois	Dallman 3	192	Wet limestone
Illinois	Duck Creek 1	416	Wet limestone
Illinois	Marion 4	173	Wet limestone
Illinois	Newton 1	617	Dual alkali
Illinois	Will County 1	188	Wet limestone
Indiana	A. B. Brown 1	265	Double alkali
Indiana	Dean H. Mitchell 11	115	Wellman Lord
Indiana	Gibson 5	668	Wet limestone
Indiana	Merom 1—2	980	Wet limestone
Indiana	Petersburg 3	531	Wet limestone
Indiana	R. M. Schahfer 17	393	Dual alkali
Kansas	Jeffrey 1—3	2160	Wet limestone
Kansas	La Cygne 1	873	Wet limestone
Kansas	Lawrence 4—5	517	Wet limestone
Kentucky	Cane Rune 4—6	644	Wet limestone/dual alkali
Kentucky	D. B. Wilson 1	440	Wet limestone
Kentucky	East Bend 1—2	1296	Wet limestone
Kentucky	Green River 1—2	75	Wet limestone
Kentucky	Green 1—2	480	Wet limestone
Kentucky	H. L. Spurlock 2	508	Wet limestone
Kentucky	Mill Creek 1—4	1702	Wet carb. limestone
Kentucky	Paddys Run 6	75	Wet carb. limestone
Kentucky	Paradise 1—2	1408	Wet limestone
Michigan	J. B. Sims—3	81	Wet limestone
Michigan	Michigan Project 1	55	Wet limestone
Michigan	Shiras 3	43	Dry limestone
Mississippi	Morrow 1—2	406	Wet limestone
Minnestoa	Clay Boswell 3—4	918	Wet limestone/alk fa
Minnesota	Riverside 1—2,6	138	Dry limestone
Minnesota	Sherburne County 1—2	1440	Wet limestone
Missouri	Hawthorn 3—4	255	Wet limestone
Missouri	Sikeston 1	235	Wet limestone
Missouri	Southwest 1	194	Wet limestone
Missouri	Thomas Hill 3	730	Wet limestone
Montana	Coal Strip 1—3	1494	Wet limestone
North Dakota	Antelope Valley 1	455	Dry limestone
North Dakota	Coal Creek 1—2	1100	Wet limestone/alk fa
North Dakota	M. R. Young 2	440	Wet limestone
North Dakota	Coyote 1	456	Dry aq carb.
Nevada	Reid Gardner 1—4	616	Wet sodium carb.
New Mexico	Four Corners 1—3	573	Wet limestone/alk fa
New Mexico	San Juan 1—4	1708	Wellman Lord
New Mexico	Plains 1	233	Wet limestone
New York	Somerset 1	625	Wet limestone
Ohio	Conesville 5—6	750	Wet limestone
Pennsylvania	Bruce Mansfield 1—3	2505	Wet limestone
Pennsylvania	Cromby 1	187	Magnesium oxide
Pennsylvania	Eddystone 1—2	707	Magnesium oxide
Pennsylvania	Elrama 1—4	510	Wet limestone

APPENDIX B (continued)
List of Plants and Type of FGD System Operations in 1984[4]

State	Plant name	Capacity [MW(e)]	FGD system type
Pennsylvania	F. R. Phillips 1—4	411	Wet limestone
Pennsylvania	Mitchell 33	—	Wet limestone
South Carolina	Winyah 2—4	945	Wet limestone
Texas	Martin Lake 1—3	2379	Wet limestone
Texas	Monticello 3	793	Wet limestone
Texas	Gibbons Creek 1	443	Wet limestone
Texas	San Miguel 1	447	Wet limestone
Texas	Sandow 4	591	Wet limestone
Texas	W. A. Parrish 8	614	Wet limestone
Utah	Hunter 1—3	1323	Wet limestone
Utah	Huntington 1	446	Wet limestone
West Virginia	Pleasants 1—2	1368	Wet limestone
Wyoming	Jim Bridger 4	508	Wet sodium carb.
Wyoming	Laramie River 1—3		
Wyoming	Naughton 3	326	Wet sodium carb.

REFERENCES

1. Energy Information Administration, U.S. Department of Energy, *Electric Power Monthly,* 1985.
2. National Coal Ash Association, Inc., *Regional Ash Production and Utilization: Summary,* Washington, D.C., 1984.
3. Envirosphere Company, Economic analysis of impact of RCRA on coal combustion by-products utilization, Appendix G in *Report and Technical Studies on the Disposal and Utilization of Fossil-Fuel Combustion By-Products,* Edison Electric Institute, Washington, D.C., 1982.
4. *Power Directory 1984/1985: An Environmental Directory of U.S. Steam Electric Power Plants,* Edison Electric Institute, Washington, D.C., 1984, 387.
5. Abernathy, R. F., et al., *Major Ash Constituents in U.S. Coals,* Bureau of Mines, Rep. No. 7240, U.S. Government Printing Office, Washington, D.C., 1969.
6. Rudy, J. D. and Huettenhain, H., Western Subbituminous Coals and Lignites, Electric Power Research Institute, Rep. No. CS-1768, Palo Alto, Calif., 1981.
7. O'Gorman, J. V. and Walker, P. L., Mineral Matter and Trace Elements in U.S. Coals, prepared by Coal Research Section, Pennsylvania State University for the Office of Coal Research, U.S. Department of Interior, Washington, D.C., R&D Rep. No. 61, 1972.
8. Kurgan, G. J., Balestrino, J. M., and Daley, J. R., *Coal Combustion By-Products Utilization Manual, Volume 1: Evaluating the Utilization Option,* Electric Power Research Institute Rep. No. CS-3122, Palo Alto, Calif., 1984, 698.
9. Soholt, L. F., et al., *Handling of Combustion and Emission Abatement Wastes from Coal-Fired Power Plants,* U.S. Department of Interior, Washington, D.C., FWS/OBS-80/33, 1980.
10. Munro, L. A., *Chemistry in Engineering,* Prentice-Hall, Englewood Cliffs, N.J., 1964.
11. Singer, J. G., Ed., *Combustion Fossil Power Sytems,* 3rd ed., Combustion Engineering, Windsor, Conn., 1981.
12. Summers, K. V., Rupp, G. L., and Gherini, S. A., *Physical-Chemical Characteristics of Utility Solid Wastes,* Electric Power Research Institute Rep. No. EA-3236, Palo Alto, Calif., 1983.
13. McIlvaine Company, *U. S. Utility Particulate Control,* June 1979.
14. Oglesby, S., Jr. and Nichols, G., *Particulate Control Highlights,* U.S. Environmental Protection Agency, Washington, D.C., EPA-600/8-77-020a, 1977.
15. Wark, K. and Warner, C. F., *Air Pollution: Its Origin and Control,* 2nd ed., Harper & Row, New York, 1981.
16. Controlling Particulate Emissions from Coal-Fired Boilers, U.S. Environmental Protection Agency, Washington, D.C., EPA-600/8-79-016, 1979.

17. Behrens, G. P., Jones, G. D., Meserole, N. P., and Seams, W. S., *The Evaluation and Status of Flue Gas Desulfurization Systems,* Electric Power Research Institute Rep. No. CS-3322, Palo Alto, Calif., 1984.
18. Knight, R. G. and Rothfuss, E. H., *FGD By-Product Disposal Manual,* 3rd ed., Electric Power Research Institute, Rep. No. CS-2801, Palo Alto, Calif., 1983, 884.
19. Envirosphere Company, Information responding to EPA's request regarding burning and cotreatment/codisposal of low-volume wastes generated at fossil fuel-fired electric generating stations, prepared for Utility Solid Waste Activities Group and Edison Electric Institute, Washington, D.C., August 1981.
20. Energy Information Administration, U.S. Department of Energy, Cost and Quality of Fuels for Electric Utility Plants, Washington, D.C., 1979.
21. Energy Information Administration, Department of Energy, Washington, D.C., FPC Form 67, 1980.
22. National Ash Association, *Comparative Ash Collection and Utilization in the U.S., 1966 to 1983,* Washington, D.C., 1985.
23. Utility Data Institute, *1981 Generating Capacity, Coal Consumption, and Combustion By-Product Generation at Operating Coal-Fired Power Plants over 100 MWe Capacity,* Edison Electric Institute, Washington, D.C., 1983.
24. Energy Information Administration, U.S. Department of Energy, Washington, D.C., Cost and Quality of Fuels for Electric Utility Plants, DOE/EIA-0191, 1983.
25. Rose, S. G., Dane, J., Eynon, B. P., and Switzer, P., *Extraction Procedure and Utility Industry Solid Wastes,* Electric Power Research Institute, Rep. No. EA-1667, Palo Alto, Calif., 1981, 45.
26. Hulett, L. D., Weinberger, A. J., Ferguson, N. M., Northcutt, K. J., and Lyon, W. S., *Trace Element and Phase Relations in Fly Ash,* Electric Power Research Institute, Rep. No. EA-1822, Palo Alto, Calif., 1981, 63.
27. Turner, R. R., Lowry, P., Levin, M., Lindberg, S. E., and Tamura, T., *Leachability and Aqueous Speciation of Selected Trace Constituents of Coal Fly Ash,* Electric Power Research Institute, Rep. No. EA-2588, Palo Alto, Calif., 1982, 94.
28. Kuryk, B. A., Bodek, I., and Santhanam, C. J., *Leaching Studies on Utility Solid Wastes: Feasibility Experiments,* Electric Power Research Institute, Rep. No. EA-4215, Palo Alto, Calif., 1985, 316.
29. Switzer, P., Eynon, B. P., and Holcombe, L. J., *Pilot Study of Time Variability of Elemental Concentrations in Power Plant Ash,* Electric Power Research Institute, Rep. No. EA-2959, Palo Alto, Calif., 1983, 180.
30. Holcombe, L. J., Eynon, B. P., and Switzer, P., *Time Variability of Elemental Concentrations in Power Plant Ash,* Electric Power Research Institute, Rep. No. EA-3610, Palo Alto, Calif., 1984, 182.
31. Bahor, M. P., McLaren, R. J., Niece, J. E., and Pederson, H. C., *Coal Ash Disposal Manual,* 2nd ed., Electric Power Research Institute, Rep. No. CS-2049, Palo Alto, Calif., 1981, 440.
32. U.S. Federal Power Commission, *The Status of Flue Gas Desulfurization Applications in the United States: Technological Assessment,* Washington, D.C., 1977.
33. SCS Engineers, Inc., Recovery, Utilization and Disposal of Solid By-Products Generated by Dry Flue Gas Desulfurization Systems: State of the Art and Research Needs, *Electric Power Research Institute,* Rep. No. CS-1765, Palo Alto, Calif., 1981.
34. Michael Baker, Jr., Inc., *Stabilization of Power Plant By-Product Storage Sites for Building Development,* Electric Power Research Institute, Rep. No. CS-3475, Palo Alto, Calif., 1984, 196.
35. Kaakinen, J. W., Jorden, R. M., Lawasani, M. H., and West, R. E., Trace element behavior in coal-fired power plants, *Environ. Sci. Technol.,* 9(9), 862, 1975.
36. Bern, J., Residues from power generation: processing, recycling, and disposal, in *Land Application of Waste Materials,* Soil Conservation Society of America, Ankeny, Iowa, 226, 1976.
37. Seals, R. K., Moulton, L. K., and Ruth, B. E., Bottom ash: an engineering material, *J. Soil Mech. Found. Div.,* ASCE.SM4, 311, 1972.
38. Page, A. L., Elseewi, A. A., and Stranghan, I. R., Physical and chemical properties of fly ash from coal-fired power plants with reference to environmental impacts, *Residue Rev.,* 71, 83, 1979.
39. Bacon, L. D., Fly ash for construction of highway embankments, in *Proc. 4th Int. Ash Util. Symp.,* Fabor, J. H., Babcock, A. W., and Spencer, J. D., Eds., ERDA, Morgantown Energy Research Center, W. Va., MERC/SP-7614, 262, 1976.
40. Roy, W. R., Thiery, R. G., Schuller, R. M., and Suloway, J. J., *Coal fly ash: A review of the literature and proposed classification system with emphasis on environmental impacts.* Environmental Geology Notes No. 96. Illinois State Geological Survey, Champaign, Ill., 69, 1981.
41. Dudas, M. J., Long-term leachability of selected elements from fly ash, *Environ. Sci. Technol.,* 15, 840, 1981.
42. Warren, C. J. and Dudas, M. J., Weathering processes in relation to leachate properties of alkaline fly ash, *J. Environ. Qual.,* 13, 530, 1984.

43. Van der Sloot, H. A., Wijkstra, J., et al., Leaching of trace elements from coal solid waste, *Energienderzoek Centrum Nederland,* ECN-120, Petten, The Netherlands, 201, 1982.
44. Radian Corporation, Waste and Water Management for Conventional Coal Combustion: Assessment, Report-1979, U.S. Environmental Protection Agency, No. EPA 600/7-80-012e, Washington, D.C., 1980.
45. Guidelines for Procurement of Fly Ash in Cement and Concrete, U.S. Environmental Protection Agency, 40 CFR 249, Washington, D.C., 1983.
46. National Ash Association, How fly ash improves concrete block, ready-mixed concrete and concrete pipe, *Concrete Industries Yearbook,* undated.
47. Bollin, G., Dundee Cement, personal communication, January 10, 1980.
48. Fowler, G. G. and Styron, R. W., Fly ash in paving and structural concrete, in *5th Int. Ash Util. Symp.,* 1979.
49. Adriano, D. C., Page, A. L., Elseewi, A., Chang, C., and Straughan, I. R., Utilization and disposal of fly ash and other coal residues in terrestrial ecosystems: A review, *J. Environ. Qual.,* 9, 333, 1980.
50. Elseewi, A., Grimm, S. R., Page, A. L., and Straughan, I. R., Boron enrichment of plants and soils treated with coal ash, *J. Plant Nutr.,* 3, 409, 1981.
51. Mulford, F. and Martens, D. C., Response of alfalfa to boron in fly ash, in *Soil Sci. Soc. Am. Proc.,* 35(2), 296, 1971.
52. Elseewi, A. A., Bingham, F. T., and Page, A. L., Growth and mineral composition of lettuce and swiss chard grown on fly-ash amended soils, *Environmental Chemistry and Cycling Processes,* Adriano, D. C. and Brisbin, T. L., Jr., Eds., Savannah River Ecology Laboratory, NTIS CONF-760429, 1978.
53. Doran, J. and Martens, D., Molybdenum availability as influenced by application of fly ash to soil, *J. Environ. Qual.,* 1(2), 186, 1972.
54. Schnappinger, M., Martens, D., and Plank, C., Zinc availability as influenced by applications of fly ash to soil, *Environ. Sci. Technol.,* 9(3), 258, 1975.
55. Elseewi, A. A. and Page, A. L., Molybdenum enrichment of plants grown on fly ash-treated soils, *J. Environ. Qual.,* 13(3), 394, 1984.
56. Elseewi, A. A., Bingham, F. T., and Page, A. L., Availability of sulfur in fly ash for plants, *J. Environ. Qual.,* 7, 69, 1978.
57. Buckman, H. and Brady, R., *The Nature and Properties of Soils,* Macmillan, New York, 1969.
58. Reneau, R., Department of Agronomy, School of Agriculture and Life Sciences, Virginia Polytechnic Institute and State University, Blacksburg, telephone conversation, February 21, 1980.
59. Phung, H., Lund, L., and Page, A. L., Potential use of fly ash as a liming material, *Environmental Chemistry and Cycling Processes,* Adriano, D. C. and Brisbin, T. L., Eds., Savannah River Ecology Laboratory, NTIS CONF-760429, 1978.
60. Reisenhauer, H., Walsh, L., and Hoeft, R., Testing soils for sulfur, boron, molybdenum, and chlorine, *Soil Testing and Plant Analysis,* Walsh, L. and Beaton, J., Eds., Soil Science Society of America, Madison, Wis., 1973.
61. Brown, J. and Snell, P., The use of PFA in land reclamation, in *1st Int. Conf. Ash Technol. Mark.,* London, Paper 2.3, 1978.
62. Walker, W., Kussow, W., Schulte, E., Horn, M., and Delwiche, P., *Liming Agricultural Soils With Fly Ash: Preliminary Field Evaluation,* Modern Waste Conference, 1979.
63. Capp, J. P., United States Department of the Interior Bureau of Mines, Morgantown, W.Va., personal communication, January 16, 1980.
64. Utility Solid Waste Activities Group, petition for a federal procurement guideline for the use of power plant ash, submitted to U.S. Environmental Protection Agency, Washington, D.C., 1984.
65. Barengerg, E. J., A state of the art on lime-fly ash stabilized base and subbase courses, Ash Short Course, Arizona State University, Tempe, Ariz., 1978.
66. The Port of New York Authority, *Newark Airport Redevelopment: The Pavement Story,* Port Authority, New York, 1968.
67. Kinder, D. L., Cement — stabilized bottom ash and subbase courses, in *Proc. Technol. Util. Power Plant Ash Symp.,* Tempe, Ariz., 1978.
68. Grancher, R. A., The lightweight aggregate boom, *Rock Products,* December, 1969.
69. Lytag Co., A unique manufacturing process for lightweight aggregates, London, 1978.
70. Casamatta, D., Penn and Virginia Materials, telephone conversation, January 15, 1980.
71. Rowen, A. R., Utilization experience in Detroit, Fourth Annual Technical Meeting, National Ash Association, 1973.
72. Roy, N. K., Murtha, M. J., and Burnet, G., Recovery of iron oxide from power plant fly ash by magnetic separation, paper presented at International Conference on Industrial Applications of Magnetic Separations, Rindge, N.H., 1978.

73. Roy, N. K., Murtha, M. J., and Burnet, G., Use of the Magnetic Fraction of Fly Ash as a Heavy Medium Material in Coal Washing, Ames Laboratories, U.S. Department of Energy and Dept. of Chemical Engineering, Iowa State University, unpublished, 1979.
74. Canon, R. M., Seeley, F. G., DeCarlo, V. A., McDowell, W. J., Watson, J. S., and Brown, K. B., Removal and recovery of metals from fly ash, in *Proc. Conf. Ash Technol. Mark.*, London, 1979.
75. Burnet, G. and Murtha, M., Direct utilization-recovery of minerals from coal fly ash, *Tech. Prog. Rep.*, 3rd Quarter, 1979, Ames Laboratory, Ames, Iowa, 1980.
76. New York State Energy Research and Development Authority, *Coal-Waste Artificial Reef Program*, Electric Power Research Institute, Report No. CS-3936, Palo Alto, Calif., 132, 1985.

INDEX

A

B

C

D

E

F

G

H

I

K

L

M

N

O

P

Q

R

S

T

U

V

W

Z